Puffins

Euan Dunn

BLOOMSBURY
LONDON • NEW DELHI • NEW YORK • SYDNEY

The RSPB is the country's largest nature conservation charity, inspiring everyone to give nature a home so that birds and wildlife can thrive again.

By buying this book you are helping to fund The RSPB's conservation work.

If you would like to know more about The RSPB, visit the website
at www.rspb.org.uk
or write to: The RSPB, The Lodge, Sandy, Bedfordshire,
SG19 2DL; 01767 680551.

First published in 2014

Bloomsbury Publishing Plc, 50 Bedford Square, London WC1B 3DP
Bloomsbury USA, 175 Fifth Avenue, New York, NY 10010

www.bloomsbury.com
www.bloomsburyusa.com

Bloomsbury Publishing, London, New Delhi, New York and Sydney

A CIP catalogue record for this book is available from the British Library
Library of Congress Cataloging-in-Publication Data has been applied for

Commissioning editor: Julie Bailey
Project editor: Jasmine Parker
Series design: Rod Teasdale
Typesetting and layout by Susan McIntyre

ISBN (print) 978-1-4729-0354-9
ISBN (ebook) 978-1-4729-0355-6

Printed in China by C&C Offset Printing Co Ltd.

10 9 8 7 6 5 4 3 2 1

Contents

Meet the Puffins

If you ask someone to name any British seabird, or indeed any British bird at all, the chances are that 'Puffin' will be among the first names to spring to their lips. The Puffin is one of our best-loved and best-known birds, amounting almost to a national treasure. It adorns children's story books, has graced postage stamps and is the photogenic pin-up bird of magazines – so much so that over the years it has featured on the front cover of the RSPB's magazine more often than any other species.

As a family, seabirds in general are a sober-looking bunch, but the Puffin brings a touch of the exotic to our shores. With its garish striped beak, wistful eyes and orange feet, it is the charismatic clown of the seabird world, while its rolling upright gait and dapper dinner-suit plumage invite comparison with penguins and no doubt also with the whimsy in ourselves.

For whatever reason, the Puffin strikes a unique chord with us. It is this empathy that continues to make it by far the biggest star attraction for visitors to seabird islands and reserves around our coasts. So popular and iconic is the Puffin that for many it has become a bellwether for the health of our seas; if Puffins are doing well, perhaps all is well with our coastal waters. On the other hand, if the Puffin is in trouble, our alarm bells cannot help but ring more loudly than if the victim were, say, a gull or a Common Guillemot. In order to gain insight into the state of 'puffindom' and its place in the marine environment, however, we need first to lay bare the daily life of the Puffin and explore how it is adapted to spend most of the year offshore in the storm-tossed ocean and the rest of its existence onshore in the social whirl of a colony.

Above: There is something in the Puffin's look that chimes with ourselves.

Opposite: With its unique rainbow beak, the Puffin is unmistakable in the seabird clan.

The Puffin's cousins

Puffins belong to the family of 22 species of seabird known as auks, a group of compact, pigeon-sized birds thriving on a diet of small fish and crustaceans, which are chased underwater and caught by the birds. Though they undertake extensive seasonal migrations, Puffins and indeed all their fellow auks are confined to the northern hemisphere, where they fill the same niche in the marine environment as penguins do in the southern hemisphere.

Global distribution apart, penguins differ from Puffins and other auks in one other key respect – they are all flightless, whereas all auks can fly. The only exception to this is the Great Auk, which paid the ultimate price for its flightlessness by making itself an easy target for man. The magnificent Great Auk was hunted to extinction in the 19th century, the last known British individual having been killed on the remote island of St Kilda in 1840. Four years later the last two known survivors of the species met the same fate on Iceland.

There are four species of puffins, three of which live in the North Pacific. The species that is found in the UK and continental Europe, and the smallest of the four, is confined to a broad swathe of ocean straddling the North Atlantic from New England (USA) in the west to Novaya Zemlya (Russia) in the east, numbering in total an estimated 20 million individuals. To distinguish it from its Pacific relatives, our Puffin is formally known as the Atlantic Puffin (*Fratercula arctica*), but for simplicity in this book we mostly just call it the Puffin.

Below: Our Puffin is formally known as the Atlantic Puffin.

Pacific Puffins

Above: The Horned Puffin.

Above: The Tufted Puffin.

Above: The Rhinoceros Auklet.

The names of the three Pacific puffin species all reflect their dramatic facial decoration in the breeding season. The Horned Puffin (*Fratercula corniculata*) most closely resembles our Atlantic Puffin, but it has a largely yellow bill and a small fleshy 'horn' projecting above each of its eyes. The Tufted Puffin (*Fratercula cirrhata*) is the largest of the four, with a massive orange bill, long, straw-coloured plumes sweeping back from its crown and a blackish body. Lastly, and least Puffin-like of all, the Rhinoceros Auklet (*Cerorhinca monocerata*) is so-called due to the pale, horn-like knob sticking up from the base of its upper bill, but this puffin lacks the spade-like conical bill of the other three species and also has much drabber, sooty-brown plumage.

A rare vagrant

The Pacific trio of puffins wander widely within their own oceans but, as far as the records show, not beyond them. So the sighting of a Tufted Puffin (*Fratercula cirrhata*) – the first ever record for Britain – on the north Kent coast in September 2009 was a jaw-dropping moment for the few birdwatchers who had the great good fortune to spot and photograph it. Perhaps the bird was swept across the Atlantic by a gale. Alternatively, climate change may have played a role in its initial escape from the Pacific to the Atlantic, given that melting ice in the Arctic is increasingly opening up a sea corridor north of Canada where previously there was an impenetrable frozen barrier to the free passage of seabirds.

The gaudy sea parrot

Like its Pacific relatives, in adulthood our Puffin possesses remarkable facial adornments in summer. The centrepiece is a conical beak of brick-red, blue and cream, much deeper than it is wide, and featuring curved, vertical grooves (see also page 34), an effect that earned the bird its ancient name of 'sea parrot'. On either corner of the base of the beak nestles a fleshy saffron-yellow rosette, its wrinkling making it look a bit like a walnut kernel. Each dark eye is encircled by a crimson ring, above and below which are small, bluish plates, the triangular upper one lending the Puffin a look that hovers between wistful and a quizzical 'who me?' The visual impact of all these decorative effects is concentrated by a backdrop of greyish-white cheeks.

Below: The base of the beak is flanked on either side by a distinctive fleshy rosette.

In addition to the show of colour on the head, the legs and feet are an equally eye-catching orange. All this colour intensity is heightened by the monochrome plumage covering the rest of the adult Puffin's body, a dazzling white breast contrasting with black upperwings, back and tail. The black sweeps up over the crown of the head to the base of the beak like a hood. It is this 'friar's cowl' feature that gave rise to the Puffin's scientific name, *Fratercula arctica,* meaning 'little brother of the Arctic'. The Puffin's portly, barrel-chested outline only serves to reinforce this friar-like impression.

The gaudy clown's mask is worn only during the breeding season. After its summer exertions are over, the Puffin undergoes a dramatic winter makeover unrivalled among seabirds. Like a knight quitting the field of battle, the Puffin dismantles its finery, just as if it were moulting feathers, and melts into the open ocean to lie low until spring signals that it is time to gird up again for the joust.

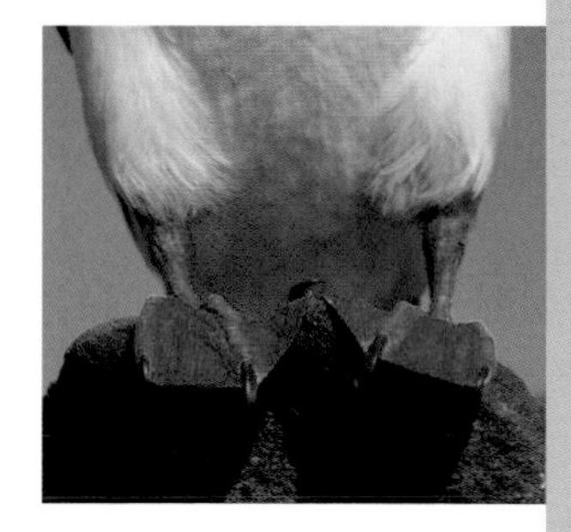

Above: The large, webbed feet are an eye-catching orange-yellow.

Top: The Puffin's gaudy clown's mask is worn only during the breeding season.

Above: With its darker and more washed-out appearance, the Puffin in winter was once considered a separate species.

The heraldic beak turns out to be only a superficial facade, its bright horny plates being shed in winter to reveal an altogether more sombre dark foundation. The eye plates also drop off, while the rosettes at the base of the beak shrivel and fade. To complete this facial transformation, the off-white cheeks darken to a dusky grey. Even the orange legs and feet transmute to a dull yellow. So radical is this seasonal switch of colour and structure that Puffins in winter were once thought to be a separate species.

In summer a minority of Puffins arrive back at the colony still in winter plumage and in no condition to breed. On Skomer Island (West Wales), several such birds return every year, but always in late May or early June when the season is well advanced. One winter-plumage bird has returned to the same spot on Skomer for the last nine years, a puzzlingly bankrupt pattern, since this Puffin seems condemned to forfeit the chance to breed year after year.

Right: This Puffin is out of kilter with the seasons, returning summer after summer to Skomer Island in winter plumage.

Snow-white Puffin

Very rarely, a genetic mutation throws up a freak – a Puffin with almost all-white plumage. Such birds with a dilution of pigment are called 'leucistic'. They are not 'albino' as they do not have red eyes but, much more obviously, their beak and eye ornaments are the same colour as those of normal Puffins. One such specimen is on display in the Natural History Museum in London. In olden times white Puffins had legendary status among superstitious sailors and in the folklore of remote island communities that harvested Puffins (as some still do, see pages 90–107) and other seabirds for food. A celebrated white Puffin was said to have come back to the Faroe Islands every spring for 50 years which, as we will see, is not quite as mythical as it sounds.

Above: Specimens of white Puffin, like this one in London's Natural History Museum, were prized by collectors.

Below: A rare 'leucistic' Puffin photographed off the Isles of Scilly in 2010.

Size matters

Leaving aside the extremes of white Puffins, regular Puffins also vary in one significant dimension, namely their size and weight. Puffins that live in the Arctic tend to be bigger than those further south, a general trend in animals known as Bergmann's rule. In 1847 Carl Bergmann, a German biologist, explained this on the grounds that bigger animals have a lower surface area to volume ratio than smaller animals, and therefore – weight for weight – they radiate less body heat. This, Bergmann theorised, enabled the relatively larger individuals in northern latitudes to keep warmer in a colder climate. In this geographical spectrum, Puffins that live in the UK and Ireland are among the smallest, weighing in at about 400g (14oz). In the high Arctic, by contrast, Puffins in Spitsbergen (Svalbard) and north-east Greenland are some 200g (7oz) heavier on average than ours and have a truly massive spade of a bill to match. The difference between them is so great that it is still a matter of debate whether the supersized cousin of our Puffin is actually a separate subspecies of the Atlantic Puffin.

Right: In Spitsbergen and elsewhere in the far north of their range, Puffins are supersized and sport a massive beak.

Above: Puffins that live in the UK and Ireland are much smaller than their Arctic cousins.

Even within our UK Puffin population, there is further size variation. As in most seabirds, there are no plumage differences to distinguish the sexes, so outwardly adult male and female Puffins look very similar. On average, however, males are slightly larger (including in beak size) and heavier than females. Unfortunately, given the amount of size overlap, this will not help you to tell males and females apart by eye at a Puffin colony.

To add a final twist, Puffins do not stay the same weight throughout the breeding season. Providing they have not suffered food shortages in winter (and that is by no means guaranteed these days), Puffins are generally heaviest when they first return to the colony in spring. However, they shed calories during the course of breeding, most likely due to the sheer effort involved in raising a chick, and are generally lighter by the time their offspring fledge.

Puffin vocalisations

Above: The Puffin makes use of a distinctive voice but, unlike many other seabirds, it isn't raucous and doesn't carry around the island.

Below: When bill-gaping to stand its ground against a rival, the Puffin gives a call best described as a creaking or gurgling sound.

We think of most seabirds as noisy birds, especially at a breeding colony. Gulls are noted for their caterwauling, terns for their piercing screams, Northern Gannets for their gurgling roar, and Manx Shearwaters for their outlandish banshee calls. Some auk species are equally vociferous, with Common Guillemot and Razorbill colonies being a cacophony of crowing and guttural rattles which – with the right acoustics – can deafen like a motorbike rally.

Puffins, by contrast, are relatively quiet birds – their vocal repertoire is confined largely to throaty groans, growls, creaks and nasal moaning sounds. Many of these come across as of a subdued and intimate mate-to-mate nature – it is almost impossible to put them into words, far less to interpret what they are signalling. As such, it is impossible to define the telltale call of a Puffin with the precision of, for example, the diagnostic 'kitti-wake' cry of the Kittiwake.

Deeper is better

It is a well-known fact that in dense habitats where sound does not carry well, the voices of animals that penetrate best are deep and resonant – that is why rainforest birds characteristically have just such voices. It is therefore possible that the generally low frequency of the Puffin's vocal range (mostly around 1 kilohertz, compared with up to 5–6 kilohertz in the cliff-nesting Common Guillemot) is an adaptation to the need for their most critical exchanges to be conducted underground in breeding burrows. Kenny Taylor (1993) has described this eloquently: 'It is a strange experience to stand in a puffinry where no puffins are visible above ground and hear the groaning calls of many birds from the burrows around you, as if the earth itself was speaking in the tongues of a hundred puffins.' There is no evidence that in the Puffin the calls differ between the sexes, but this is an area in need of more research. In so far as we can ascribe meaning to it, we will come back to the Puffin's use of voice as a communication tool in the next chapter.

Below: The low pitch of the Puffin's calls may be an adaptation for better sound transmission underground.

Colonial Life

Like the vast majority of seabird species, Puffins are highly gregarious in the breeding season, choosing to nest alongside one another in colonies often tens of thousands strong, on exposed cliffs and islands. A major Puffin colony, pulsing with movement, colour, sound and smell, all part of the everyday drama of raising offspring, is one of the most exhilarating assaults on the senses these islands can offer.

So why does communal breeding benefit Puffins, what constitutes a good place to breed and how do Puffins check out the colonies on offer before deciding which one is to be their home base? We get used to seeing gulls ashore all the year round, but the Puffin's life on land is but a brief summer encounter. For the rest of the year the precise whereabouts of Puffins used to be a mystery, but modern technology is transforming our understanding of their oceanic wanderings.

Left: Part of the breeding colony on the Farne Islands (Northumberland), England's biggest Puffinry.

Below: A thriving colony area on the Isle of May, the premier breeding site for Puffins on Scotland's east coast.

Why Puffins stick together

Above: The island of Dun in the St Kilda archipelago, the UK's greatest Puffin stronghold.

No-one knows how many Puffins there are in the world, but the best estimate is around 20 million birds. Most live in the north of their breeding range, with Iceland supporting some three million pairs, around half that number inhabiting the next most important stronghold of Norway, followed by the Faroe Islands with fewer than half a million pairs. About half of all Iceland Puffins breed on the Westmann Islands, which support the largest colonies in the world. In Britain and Ireland there are probably around half a million pairs, distributed in a halo of breeding colonies girdling our coast. The biggest of these is in St Kilda, an island citadel and our greatest seabird metropolis situated about 60km (37 miles) west of the Outer Hebrides in north-west Scotland, and home to perhaps a third of the UK breeding population. The chapter 'Watching Puffins' describes the best places to see Puffins, but apart from in the Outer Hebrides other major concentrations are found on Orkney and Shetland, and in eastern Scotland, north-east England and the Pembrokeshire islands in Wales.

In these favoured places, Puffins typically congregate to breed in large colonies, so much so that in St Kilda in the 19th century, the minister on the island, the Rev. Neil MacKenzie, likened them to a 'small cloud of locusts' so dense as to cast a dark shadow over the ground as they wheeled overhead. So Puffins are highly social in the breeding season, a pattern found in most other seabird species.

Why do Puffins nest in colonies at all? The first known fossil of a Puffin-like bird was found in deposits five million years old in a North Carolina (USA) phosphate mine. How Puffins lived then will forever remain a mystery, but clearly, to have evolved to be so gregarious over the millennia since then, living in groups must have conferred a powerful selective advantage over the solitary life.

Biologists have speculated long and hard about just what benefits derive from the colonial life for Puffins and other seabirds. In all likelihood it is a mix of things. For a start, a colony is a kind of dating agency or club where Puffins can meet others of their own kind in abundance and enjoy multiple choice in seeking a suitable mate. Although all Puffins may look almost identical to us, they must see one another very differently and there is

Above: Puffins sometimes congregate in such numbers that they have been likened to a cloud of locusts when they take to the air.

Below: Puffins in flight over the Isle of May where the 2013 census yielded 46,000 occupied burrows.

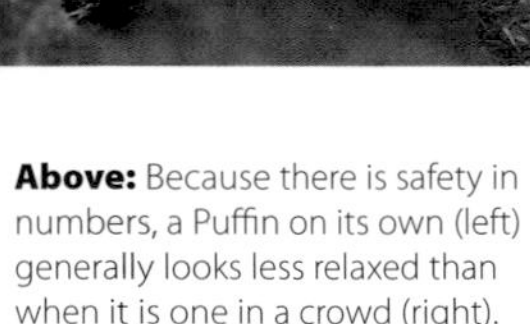

Above: Because there is safety in numbers, a Puffin on its own (left) generally looks less relaxed than when it is one in a crowd (right).

growing evidence to show that birds are as choosy about mates as we are. For the Puffin, the priority is to select an individual with good genes who will be successful at raising offspring.

In addition, Puffins face threats from a variety of predators, and simple game theory predicts that, for the individual Puffin, its risk of being singled out by a predator is reduced if it is surrounded by others of its kind, a phenomenon known in animal behaviour as 'the selfish group'. By bunching in tight shoals (commonly called bait balls), the individual small fish that Puffins hunt underwater are adopting the very same survival strategy. All the birds in a colony of Puffins nest in the same place and at the same time (the more synchronised the better), and being in a colony is safer for a pair of Puffins than if they nested in isolation from their neighbours.

Colonial life may offer one further benefit related to the Puffin's need to search out patchily distributed fish shoals hidden under the sea, often at considerable range from the colony. A Puffin setting out from a colony on a fishing trip seems to us to have a bewildering choice of compass bearings. The suggestion is that the Puffin discovers and homes in on the location of the best feeding grounds by following whichever flight lines attract the densest Puffin traffic. Only by aggregating in a colony could Puffins generate such social cues for finding fish-rich hotspots.

It is very difficult to prove whether colonies really do function as 'information centres' like this. It is true that they often set off and return in groups, but again this may have more to do with safety in numbers than with information transfer about feeding hotspots. However, Puffins certainly do spend a lot of time sitting on headlands watching others setting off to forage or returning with fish, a degree of attentiveness that extends to all areas of colony life.

You only have to watch a group of burrows to appreciate that a Puffin colony is really a buzzing social network. The occupants of a sub-colony, or at least a nexus of burrow owners within it, behave rather like nosey residents of a tenement or a high-rise block, eager for gossip and to check out what their nearest neighbours are up to. Puffins are forever visiting and peering into the burrows of their neighbours, who seem quite tolerant of such cold calling, at best greeting the intruder with a groaning 'I see you there' call. Visiting burrows is especially prevalent in the evening, when the colony tempo slows down enough to allow what we might

Below: The residents in a Puffin colony rub shoulders like tenants in a communal block of flats.

call Puffin 'promenading', not just by solitary Puffins but also by established pairs. Any courtship or fighting between a pair becomes a public spectacle for Puffins in the vicinity, readily attracting an audience. This often has a ripple effect, sparking off other social interactions and 'conversations' nearby, so that a mounting babble of excitement seems to grip the whole burrow complex.

Such intense neighbourliness is scarcely likely to be idle inquisitiveness or voyeurism, as if Puffins were seeking vicarious thrills. Just as humans keep an eye on those around them in a block of flats, to the birds also, information is power and helps them to calibrate and adjust their own behaviour. A promenading Puffin registers a host of cues about which burrows are occupied and which ones are empty, and who is 'in' for the night or (at the start of colony occupation) choosing to roost offshore. The visitor can also glean particularly valuable information about whether the residents are paired or solitary, what stage others are at in the nesting cycle and perhaps even what fish they have been eating. Knowledge that a resident alpha male has lost his mate (or that she is late in arriving back at the start of the season) might, for example, be vital intelligence for a solitary female seeking a pairing opportunity. All of this information transfer is therefore much more than just social glue – rather it contributes in a real way to the day-to-day survival and success of the colony's inhabitants.

Below: The Puffin is intensely curious about its neighbours and forever checking out their burrows.

Settling on the right colony

Above: The Isle of May meets the Puffin's essential requirements – an island free of mammalian predators and an abundant food supply close-by.

Having evolved to breed communally, the Puffin's next challenge is where best to do it. A Puffin colony has to meet two fundamental requirements: a safe place to breed and an abundant and accessible supply of fish. Because Puffins nest on the ground, in fact underground, successful breeding demands that the colony be free from ground predators, especially rats, which play havoc with Puffin eggs and chicks. For a rat, a smelly labyrinth of Puffin burrows is like a delightful smorgasbord. The optimum formula for a successful Puffin colony is therefore a rat-free island. However, over time many formerly safe havens for Puffins have become infested with rats from shipwrecks, spelling disaster for the defenceless birds.

For example, Ailsa Craig, a turf-capped thumb of rock off the Ayrshire coast of Scotland, was said to have hosted at least a quarter of a million pairs of Puffins in the 1860s. No one knows exactly how or when Brown Rats arrived there. They may have made a tightrope scramble or swum ashore from a shipwreck, or perhaps made landfall after stowing away in the boat that plied backwards and forwards to the island, supplying coal to the lighthouse. In any event, by a century later the rats had wiped out

Above: Rats drove Puffins off Ailsa Craig, only for the birds to re-colonise when the island was rendered rat-free.

Above: The Brown Rat is the scourge of Puffin colonies, pillaging eggs and chicks from burrows.

Ailsa Craig's once-flourishing Puffin colony. In the UK the same fate befell Lundy, Puffin Island, Ramsey Island and many other islands. Fortunately this is not an irreversible state of affairs; with painstaking effort, some rat-infested islands have been rehabilitated and made safe again for Puffins to colonise and breed (see page 109).

While offshore islands are the preferred breeding places for the Puffin, a grassy cliff on the mainland is the next-best safeguard against ground predators. In Scotland there are several such mainland sites (see pages 120–123), but in England the only mainland colony of note is at the RSPB reserve of Bempton Cliffs in Yorkshire, where around 1,000 Puffins breed in the precipitous upper slopes. Bempton's topmost burrow-nesters do, however, fall victim to stoats and weasels. In the far north of the Puffin's range, exposed to the rigours of an Arctic climate, vegetation is naturally in short supply. Here, for a safe breeding refuge, the Puffin often has to resort to frost-shattered crevices in rocky cliffs and boulder fields.

From the fact that we can find the same ringed adults in the colony year after year, it follows that Puffins are mostly faithful to their chosen breeding colony. This pattern, which is typical for most seabird species, has obvious survival value in the cumulative lore each Puffin tucks away over the years about the topography of the colony, the location of the best feeding grounds and oceanographic changes in the surrounding waters, as well as a host of other factors

Below: Puffins maintain a foothold on Bempton's limestone cliffs despite the topmost burrows being vulnerable to terrestrial predators.

that could tip the balance between success or failure, including awareness of who your potential predators might be and even the vagaries of local weather patterns.

However, if all Puffins showed such colony fidelity, we would never witness the origin of new colonies or the recolonisation of formerly abandoned ones. It turns out that it is the adults that typically form this strong colony attachment, while the juveniles are the adventurous island hoppers, sampling several colonies in the gap years before they finally choose where to settle down and breed. A study of colonies on the New England coast of the USA showed that whereas juveniles may settle in the colony where they were born, they more often opt for a colony elsewhere.

On average, a juvenile's summer stopover at any one colony only lasted about two or three days, perhaps long enough to 'case the joint', but older youngsters spent longer in each colony, no doubt to refine the reconnoitring needed to help them decide where they would finally put down roots. What cues might Puffins use to reach this decision? A study of young Common Terns prospecting for a colony gives us a clue, as they were most likely to settle in a site after they had witnessed the residents experiencing high breeding success. This suggests that the youngsters were 'impressed' by, for example, witnessing adults supplying their young with an abundance of fish.

Above: Puffins prospecting for an island to join will monitor the residents to judge whether they are successful in finding food and raising young.

Below: In Iceland, where topsoil is in short supply, Puffins often nest in cliff crevices and under boulders.

Testing the waters

The wanderlust of young Puffins in search of a permanent home is known to be capable of taking them far away from their colony of birth. In a study of ringed Puffins killed by hunters on the Faroe Islands (see pages 91–107), one in every ten birds turned out to have originated in Iceland, some 800km (500 miles) to the north-west. One young Puffin colour ringed on the Isle of May in eastern Scotland also emigrated as far as the Faroes, while another from the Farne Islands in Northumberland crossed the North Sea to end up breeding in a colony in southern Norway.

However, these individuals really pushed the boundaries, and dedicated colour ringing of Isle of May Puffin chicks revealed a pattern of colony visitation closer to home. Even so, these explorations were remarkable for a scatter that spanned almost every point of the compass. With the exception of the far-flung Puffin stronghold of St Kilda, scarcely no potential UK colony was beyond the reach of these youthful voyagers from the outer Firth of Forth. Ring sightings revealed that they fanned out to visit colonies all along the east coast up to the UK's northern outpost of Shetland, and west into the Pentland Firth, Atlantic and Irish Sea colonies of Scotland, Wales and even southern Ireland.

Above: If the Puffin in the foreground should switch islands, observing its colour rings through binoculars or a telescope can reveal its colony of origin.

The colony as a brief landfall

Above: In Spring, Puffins congregate offshore from their colony before making landfall.

Left: On land Puffins gather in groups and will engage in 'loafing' or socialising in their colony.

While Puffins try to find the safest place they can to breed, the whole challenge of getting through the breeding season unscathed and raising offspring is still fraught with danger and difficulty. For this reason, Puffins are, for most of their lives, genuine 'seabirds', essentially ocean dwellers that just come ashore to breed for as little time as it takes to get the job done; they then desert the colony for the relative haven of the sea again.

On average, Puffins first appear in the waters adjacent to their UK colonies in April, although the exact timing varies across the species' range with latitude. The most southerly breeders, in Brittany and the Channel Islands, start turning up in March, but those in the far north of their range do not arrive until May. These hardy Arctic birds may start prospecting their colony even before the snow melts to reveal their familiar nesting terrain.

With breeding over, most UK Puffins leave the colony between the end of July and mid-August to resume their life on the ocean wave, so breeding is a relatively brief, four-month encounter with terra firma. For the remaining two-thirds of the year they shun the colony completely, wintering entirely offshore. Here Puffins are in their element, and – as long as they can find enough food – are quite capable of riding out most storms alone or in small groups without coming to any harm. However, we only have a hazy understanding of how and where they spend their time outside the breeding season.

Puffins from north-west Britain and Ireland fan out into the Atlantic, and some recoveries of ringed birds reveal remarkable dispersal. The metal leg rings issued by the British Trust for Ornithology carry a unique number encoding where and when a bird was ringed, and whether as an adult or chick. Ringing tells us that Puffins from north-west Britain and Ireland have undertaken journeys of epic proportions, crossing the Atlantic to Newfoundland or north to Greenland, while other individuals have wandered south as far as the Canary Islands and even into the Mediterranean. Puffins from the UK's east coast seem more conservative, largely confining themselves to the North Sea in winter, although tracking devices (see box opposite) are beginning to shed new light on this.

Above: Numbered metal Puffin rings, invaluable for showing the origin of a bird recaptured alive or found dead.

Right: An identification ring is clasped neatly with pliers round the Puffin's leg where it is no hindrance to the bird.

Plotting Puffins

Up until recently we knew nothing about the wanderings of individual Puffins outside the breeding season. However, our knowledge of where seabirds go when they leave land is being transformed with the advent of new technology in the shape of the 'geolocator', a miniature tag that logs light levels and allows us to track a bird's day-to-day movements by latitude and longitude. At the colony a geolocator – so light as to impose no extra burden on the bird – is attached to the plastic leg ring of a captured seabird. When the bird is recaptured in the following breeding season, the geolocator's light data can be downloaded and decoded.

Geolocator data from Puffins on the Isle of May in eastern Scotland showed that in early winter the birds made excursions beyond the North Sea into the Atlantic west of Scotland, a degree of outreach not previously suspected. The scientists from the Centre for Ecology and Hydrology (CEH) who made this discovery suggest that worsening winter feeding conditions in the North Sea may be forcing the Puffins further afield than in the past.

Another geolocator study of movements of Puffins from Skomer Island, led by Tim Guilford, shed remarkable light on the regularity of routes followed by particular Puffins over time. The first phase of migration, when the adult birds left the colony in August, showed a predominantly west–north-west trajectory, turning more north–north-east in autumn, and finally south in winter, with some birds flying as far as the Mediterranean, before turning back towards the colony in spring. Within this broad trend, individuals differed enormously in the directions and routes that they took, but the greatest revelation was that an individual Puffin tended to repeat the same rough track in each of the two years of the study. The scientists propose that during the years before it breeds, the Puffin learns and refines its own route, the one that 'works for it', and the oceanographic cues that mark it out. If we think of the ocean as a featureless waste, such learning challenges our imagination, but to migratory seabirds such as Puffins (as indeed to trained human seafarers) the ocean is a navigable landscape of currents, upwellings, smells, wind, sun, stars and the Earth's magnetic field, all of which are invoked as directional cues.

Above: A geolocator on the right leg of a Puffin on Skomer Island where this cutting-edge technology has opened a window into remarkable ocean wanderings.

The most recent discovery thrown up by a geolocator study is truly astonishing. At the end of their breeding season on Skellig Michael off south-west Ireland, most of the tagged Puffins travelled in about three weeks right across the Atlantic to the waters around Newfoundland and Labrador. Mark Jessopp and colleagues think the most likely explanation for this epic journey was that the birds were cashing in on the seasonal bonanza of fish there, especially capelin and sandeels. The Irish Puffins stayed in eastern Canada for just 2–6 weeks before heading back east again. Such revelations are turning our previous grasp of Puffin migration on its head.

Vital Statistics

Charles Darwin was fascinated by the way in which each species of bird, whether a woodpecker or a goose, was purpose-built for the particular 'niche' it occupied in the wild, and this appreciation of adaptation was critical in leading him to the theory of natural selection. The extreme challenge of living on exposed coasts, and plumbing the depths of the sea for small fish, has created a rare niche for the highly specialised elite of seabird species to which the Puffin belongs.

This chapter discusses some of the consequences for the Puffin of pursuing this exacting lifestyle in terms of how long it lives and takes to reach adulthood, and how changes in its beak are a clue to growing up. It is easy to think of Puffins as mindlessly following a pre-programmed lifestyle, and we even direct that jibe at unthinking humans when we call them 'birdbrained'. But for the individual Puffin, survival is a much more subtle dance than that, with a potentially long life playing a vital role in mastering a complex marine environment and passing on its genes to the next generation.

Opposite: Becoming a successful Puffin is a combination of genetic programming and hard-won experience.

Below: This adult took several years to master an extremely challenging environment.

Coming of age

Above: It will be several years before this chick is equipped with the parenting skills to enter the breeding population.

Below: A study of Sandwich Terns indicates that, as youngsters, many seabirds take months to learn how to capture fish as efficiently as their parents.

After fledging, juvenile Puffins do not start breeding until they are at least four years old and most do not do so for a few years after that, so that the average onset of breeding is at around seven years old. Why do Puffins, other auks and indeed seabird species in general need this period of adolescence or, as seabird biologists refer to it, 'deferred maturity', and why do they not just start breeding the year after they fledge? The most plausible reason for this is that it takes a significant apprenticeship to develop the physique, skills, knowledge and judgement necessary to become a truly successful and efficient Puffin.

For a start it takes time to find a suitable mate, and the quest is an elaborate ritual (see page 52). Additionally, a Puffin needs to master the art of divining where to find and how to catch highly mobile fish in the hinterland of sea around its colony. Fish of the right species and size are by no means evenly distributed and constantly available; rather they are patchily distributed and more accessible at some times of day, in some seasons and in some sea conditions than others. So a Puffin needs to learn by experience where to look for food, and once it has located it, how to catch enough of it.

Because Puffins forage deep underwater, we cannot spy on and score their youthful hunting exploits. However, by studying other seabird species we can infer the scale of the challenge facing a fishing Puffin. For example, at around six months old, Sandwich Terns (which hover before plunging headlong into the sea to pincer fish in their bills just below the surface) turn out to have a still lower rate of fish capture on average than do adults. Likewise, however much the Puffin is programmed to forage for fish, there must be a strong element of apprenticeship.

To define more precisely the challenge of what it is to become an accomplished Puffin, we also need to consider that the act of breeding requires a massive injection of additional energy. For successful breeding, a Puffin needs to find enough fish not just for itself, but also for as long as it takes to feed a hungry chick to fledging. Just as a pregnant woman is often said to be 'eating for two', the female Puffin's first challenge is to acquire enough extra nourishment to form the single egg she lays. As the egg weighs about a seventh of her body weight, egg-making is clearly not an insignificant investment of effort. For all these reasons, we can see why young Puffins take several years to learn the ropes of parenthood.

Above: A nearly fledged Puffin at its burrow entrance, taking in the outside world.

Below: A fledgling on the brink of its ocean apprenticeship exercises its wings.

The barcode beak

During its pre-breeding years, the adolescent Puffin's bill undergoes a dramatic transformation without parallel among seabirds. Over about five years it grows and evolves into the colourful, grooved, heavy-duty secateur structure typical of the mature adult. At fledging the juvenile's bill is dark, stubby and triangular in profile – altogether an unremarkable prototype of what it will become in its full glory. By the time most young Puffins first return to the colony at two years old, the beak is more substantial but still essentially triangular in shape. Although the beak is still quite dull in colour, its basic pattern has been laid down. It consists of a bluish-grey wedge of colour adjoining the head with a dark red conical bill-tip, but the surface is still relatively smooth.

Over the subsequent three years the colour brightens, the beak deepens and becomes markedly curved in outline, and two distinct grooves etch themselves on either side of the reddish forepart of the mandibles.

Above: The fledgling's beak gives little hint of the extravagant tool it will become.

Below: The two-year-old's beak already exhibits the basic architecture of the adult's.

Left: An adult beak, fully pigmented with two well-developed grooves and a less pronounced third one nearest the tip.

Below: The Razorbill has a characteristically grooved, jet black beak with a contrasting vertical white stripe.

Bottom: Of all the auk species, the Great Auk was endowed with the most bill grooves.

Thereafter most Puffins acquire a third groove and, rarely, the semblance of a fourth. With this, the Puffin's riotous sculpture of a beak is complete and is resurrected every summer for the rest of its life. We cannot stretch the 'barcode' comparison too far as the number of grooves does not enable us to nail down an individual's precise age in years. Nevertheless, if we look at a huddle of Puffins through binoculars, the mere presence or absence of grooves is a reliable guide to telling adults apart from immature birds.

Perhaps the grooves in the beak evolved to enable Puffins to sort out the sexually mature and experienced birds among those they encounter, so helping them to decide who is seriously worth courting or what the risk might be of picking a fight. Whatever their function, the grooves must be important because other auks also sport them. Thus the number of vertical grooves on the Razorbill's beak – one of the grooves is a luminous white on liquorice black – is again related to the bird's age, while the Razorbill's closest relative, the extinct Great Auk, had six to 12 such grooves on its impressive bill.

Long live the Puffin

As a general rule big bird species have a longer life expectancy than small ones. Thus the average lifespan of a Robin may be only a year or two, whereas the Wandering Albatross – the world's biggest seabird, weighing up to 12kg (26lb) and with a wingspan of 3.5m (11½ft) – has been known to live to over 60 years. Weighing no more than about 600g (21oz) and with a maximum wingspan of around 60cm (24in), the Puffin is dwarfed by the albatross, but nevertheless it shares with seabirds in general the trend towards a long lifespan.

We know how long Puffins live from repeated captures and sightings at the colonies of individuals that were first ringed as chicks. With patience, ring numbers can sometimes be read at close quarters with a telescope, which does away with the need to catch the birds. Over recent years, the record for the longest-lived Puffin has crept steadily upwards as the metal used in rings has become more durable and resistant to the corrosive effects of seawater and chafing on rocks. The relatively soft aluminium rings used a few decades ago got so worn down over time that the identification numbers engraved

Below: Ringing Puffins as chicks gives us a way of determining how old they are when checked later in life.

on them eventually became illegible or the rings even fell off, so that a ringed Puffin effectively outlived its age tag.

We now know that Puffins can live to at least 40 years of age. The oldest ever found in Europe was a 41-year-old Puffin killed by either a Raven or a Peregrine Falcon, on the Norwegian Island of Røst. We can only speculate that this individual may have ultimately died of 'natural causes' at an even riper age had it not succumbed to a predator. Long-term studies of Puffins in the UK have revealed survivors that are not far behind the Norwegian bird. The record is held by a Puffin on the Isle of May off the Firth of Forth in Scotland, which was ringed as a breeding adult in 1974 (two years after Mike Harris began to research Puffins on there, generating a uniquely deep insight into puffindom that continues to this day), and was last seen in 2008. Adding the years it must have taken this Puffin to reach breeding age, it must have been at least 39 years old when it disappeared from the colony.

While such long-lived birds are presumably near the limit of Puffin life expectancy, they are by no means unique in the colonies to which they belong. We can deduce this from looking at the recaptures of a much larger sample of the ringed population which reveals that, on average, at least nine out of ten adult Puffins typically survive from year to year. Although this varies from colony to colony, with such a high overall survival rate it is predictable that three or four decades after ringing the colony should still contain a hard core of the earliest ringed individuals.

Were it not for the evidence of ringing, it would be impossible to pick out the senior citizens in a Puffin colony since – unlike us humans – they show no outward signs of old age. Nevertheless, Mike Harris's long-term study has revealed that the likelihood of an Isle of May Puffin surviving does decline as it gets older, gradually to begin with but accelerating once it reaches the grand old age of 30. There are some indications that senile Puffins not only have less chance of surviving another year, but also may breed less well than younger Puffins. It will, however, take more years of following Puffins into advanced age to prove that this is the general rule.

Top: The Norwegian island of Røst where the oldest known Puffin to date met a premature death.

Above: Were it not for the fact it was ringed as a chick, there is no way of telling from its outward appearance how old this adult Puffin is.

Build-up to Breeding

After a long winter riding out the open ocean, spring heralds the return of Puffins to the breeding grounds, although occupation is a tentative affair and the birds do not make landfall straightaway. The ebb and flow of their convergence on the colony, the aerial spectacle of the Puffin 'wheel' (see page 41), and the intense social gatherings on the sea surface are all part and parcel of the Puffin's prenuptial ritual. In the animated sea-going rafts of birds courtship gets underway, with old pair bonds renewed and new ones forming.

Once ashore, the pace quickens as former breeders reclaim their old nesting burrows, defend them against rivals, and prepare the underground home for egg laying and chick rearing. Here the Puffin's repertoire of courtship display reaches its full expression, an exotic performance that makes maximum capital out of the Puffin's uniquely colourful body art.

Opposite: Puffins undergo a lengthy preamble before they finally settle ashore at the start of the breeding season.

Below: The significance of the Puffin's decorative face mask and beak is clear during courtship and pairing.

Homing in on land

Above: A raft of Puffins enjoy mill-pond conditions just offshore from Coquet Island (Northumberland).

Below: For each Puffin the social gathering of the raft is an icebreaker between the remoteness of their winter oceanic existence and the impending intimacy ashore.

In late winter and early spring Puffins begin to abandon their seafaring life and converge on the breeding colony. A few may venture ashore in February, but the majority do not arrive until March or early April in Britain. Since potential dangers lurk on land, it appears as though they are at first nervous about making landfall. For several days the birds are content to 'raft' in groups offshore, milling around on the sea's surface opposite the particular neighbourhood of the colony they occupied in previous years.

However, the Puffins are not just aimlessly treading water; rather this is a rendezvous for serious socialising in anticipation of breeding. By day former mates meet up here and get reacquainted, while would-be breeders practise the art of seduction, acting out displays that have their counterparts on land (see page 52). Come evening they retreat offshore again to the relative safety of the sea without having made it ashore. Eventually, however, plucking up courage, a small vanguard of Puffins makes the first foray into the colony to reconnoitre and – if they have bred there in previous years – to reclaim their former nest sites.

Typical of this overture to the breeding season is a highly fluctuating pattern of colony occupation. The first birds to settle may arrive for a few hours, only to vanish and not be seen again for a few days. As the colony swells in numbers, a cycle develops of mass attendance and absence, with peaks around every five days. Attempting to census a Puffin colony during these violent fluctuations is obviously doomed to give false readings, and the only true picture emerges from counting occupied breeding burrows. Quite why this cyclical pattern has evolved is something of a mystery, although we know that numbers ashore tend to be higher in stormy weather than in calm conditions. Intriguingly, a study of a mixed auk colony on Skokholm Island (West Wales) showed that Common Guillemots and Razorbills synchronised their attendance patterns with the Puffins, all three species peaking together.

Especially in a really big Puffin colony, days when the birds are ashore in abundance bear witness to one of the most dynamic of any seabird spectacles, which has been studied and described in exquisite detail by Puffin expert Kenny Taylor. This is the Puffin 'wheel', an aerial phenomenon in which Puffins converge on the edge of the colony in mass formation, hurtling overhead in a mighty elliptical (or less commonly figure-of-eight) path, the outer rim of the carousel carrying the air traffic out over the sea. If you sit below a major wheel, the beetling

Above: On Skokholm, the Common Guillemot (left) and Razorbill (right) synchronise their peaks and troughs of colony attendance with Puffins.

Below: A mixed colony of Puffins, Common Guillemots and Razorbills in Norway.

Right: Part of a Puffin 'wheel' at North Haven on Skomer Island, one of the great spectacles of the annual breeding circus.

fly-past of thousands of whirring wings can generate an audible muffled roar. Kenny Taylor discovered that there may be many wheels within a colony because the Puffin community attached to a particular 'sub-colony' flies over its boundaries, aligning its approach and seaward exit with distinctive features of the terrain such as ridges and gullies. Thus he mapped 21 separate wheels on one of St Kilda's islands, with the spatial array persisting from one season to the next as predictably ordered as the Olympic rings. Where sub-colonies stack in tiers like theatre balconies on a vertiginous slope, he found that the wheels follow suit with several in simultaneous motion, but not all necessarily rotating in the same direction, like in a 'spaghetti junction'.

Most seabird species just enter and leave the colony in more or less straight flight lines, so Puffin wheels – which continue throughout the breeding season – must serve some special function. The repeated circular holding pattern probably helps Puffins choose the most opportune moment to break away and land at their nest sites. The synchronised wheel may also help prevent Puffins in a crowded colony to avoid chaos and mid-air collisions with one another.

In comparison with any fluke threat from its own kind, however, the Puffin wheel surely counteracts a much

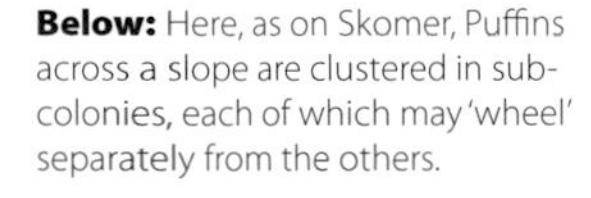

Below: Here, as on Skomer, Puffins across a slope are clustered in sub-colonies, each of which may 'wheel' separately from the others.

Peril in the air

The Puffin wheel's mutual air-traffic control mostly succeeds, but Puffins are not the most manoeuvrable of fliers; and accidents, while extremely rare, do happen. Once, while standing in the 'Cambir' sub-colony on St Kilda, I was startled by a loud thud and turned round in time to see two Puffins spinning earthwards. One lay spreadeagled on the ground, stunned but alive, while the other lay motionless. As I approached to take a closer look, the concussed bird took off, flying erratically to a distant boulder. The other had not been so fortunate and was stone dead, having clearly been pierced on the side of its face by the survivor's bill tip with such force that It suffered a depressed skull fracture. The victim had a slight bill, grey face and yellow (rather than orange) legs, indicating that it was a youngster visiting the colony. Perhaps it was inexperienced in the aerial dexterity needed for the Puffin wheel and paid the ultimate price.

Above: At times the Puffin can resemble a flying brick.

greater danger, namely that from aerial predators. Here we find another manifestation of 'the selfish group' strategy (see page 20). The maelstrom of the Puffin wheel probably bamboozles any rapacious gull or Peregrine Falcon, so reducing the risk to any one Puffin of being singled out for a meal. Moreover, the would-be predator may well think twice about the hazard to its own life and limbs before messing with the 'Hadron Collider' that is the Puffin wheel.

Below: The whirling maelstrom of the wheel may deter aerial attack by this Peregrine Falcon or by its Arctic cousin the Gyr Falcon.

Above: A relaxed loafing group of Puffins on a Skomer Island club area, bare of vegetation from constant use.

By breaking free of the wheel to make its first landfall, a Puffin signals the start of its concerted breeding effort. At the outset most Puffins do not make a beeline for the nesting area, but often seek the haven of a so-called 'club' or loafing area nearby in which they can socialise, scan the horizon in relative safety, relax and preen. Such club areas, typically a convenient rock platform or a grassy mound, are strategically placed pit stops around the perimeter of the colony, and are traditionally used year after year by succeeding generations. A club of Puffins loafing on a midsummer's day, with a fringe of pink Sea Thrift nodding in the breeze and set against a backdrop of a Prussian-blue sea, is an idyllic scene that belies the intense effort being expended at the business end of the colony.

Below: Loafing birds typically rest and preen, vital for maintaining the efficiency of their body insulation and flight feathers.

Right: In summer, Sea Thrift is as classic a backdrop to the UK Puffin colony as Bluebells are to ancient woodland.

The dating agency

Later in the season the club serves another vital function as a kind of teenage rendezvous where young, non-breeding Puffins can mingle with peers of their own and the opposite sex to develop social and courtship skills, and even seek out a suitable mate. As already mentioned, a Puffin does not breed until it is at least four years old. While experienced breeders are first to arrive at the colony in spring, most novices do not show up until later, and the younger a bird is the later it is likely to make landfall. As progressive waves of birds arrive, the colony becomes ever more crowded and busy, but the club areas are the neutral locations of choice for these novices.

While the club has a laid-back feel, it also has house rules. A Puffin dropping into the club raises its wings on landing and literally keeps a low profile. Its body language assures the other club members that it has no desire to cause trouble. Its nearest neighbours respond with a low-key raising and lowering of their feet, which seems to communicate 'OK, but over here is my spot'. We will come back to this display again, in the context of full-blown courtship display (see page 55).

Above and below: There is an etiquette to joining a club, the alighting bird (above) crouching submissively low (below) on landing.

Below: An elevated, sun-warmed slab of rock is often a preferred club area, its surface smoothed by centuries of pattering Puffin feet.

Going underground

Above: A Puffin emerges from its burrow, which it will relocate and repossess year after year.

If the Puffin's first line of defence for safe breeding is to seek an island colony free from ground predators, its second is to nest underground, using burrows in peaty soil or deep crevices among boulder fields found on sea cliffs and slopes. Puffins that have successfully nested before generally seek out their former burrows and mate up again. There is no evidence that pairs stay in close contact outside the breeding season, so – if they have not renewed their acquaintance in a raft or a club area – the burrow serves as a rendezvous where mates can meet up again at the start of the breeding season.

There are several implications of this. For a start, we should not overlook such remarkable homing skill after several months of absence on the wide ocean. Like many other long-lived birds, Puffins clearly have inbuilt 'satellite navigation' to enable them to relocate the few square centimetres of their own burrow entrance among possibly thousands of others on an expansive island slope or chaotic boulder field teeming with other Puffins.

Moreover, it is clear that Puffins aspire to reclaim the same mate and nest site from year to year, providing both have survived the rigours of winter. Puffins are thus generally monogamous for as long as both members of a pair survive. If a Puffin's mate fails to return to the colony, the survivor will generally stick with its original burrow and seek to attract a new partner.

However, as Ruth Ashcroft found on Skomer Island, something in between sometimes happens. She discovered cases in which, at the start of the breeding season, the female did eventually return but significantly later than her mate. In this situation the male hedged his bets by pairing up temporarily with a new female until his rightful mate showed up, demonstrating again the Puffin's high degree of fidelity to his familiar, tried-and-tested partner and his reluctance to divorce her. Other seabirds in which monogamy is the norm are not always so forgiving: a study of Common Terns showed that a mate was willing to wait for its delayed partner to return for about five days. After that it cut its losses and sought a replacement, which it then remained faithful to even if its original mate finally put in an appearance.

Below: Puffins are typically faithful to their previous mate, the burrow serving as a meeting place where partners can reunite after their long winter separation offshore.

Above: Each webbed foot is unique in having a distinctively angled inner claw, a tiny pickaxe to help excavate the burrow.

Where they co-exist with Rabbits or Manx Shearwaters, Puffins sometimes commandeer ready-made burrows, but more often the Puffin digs its own. Both sexes take part in excavating, but the male generally takes on the lion's share of the work. Here the Puffin's beak proves to be not only for show, but also a formidable digging tool. Now we also see the adaptive value of the muscular bull neck for making powerful, delving stabs into the turf, which can present quite a dense barrier, having been fertilised by decades of droppings rich in nitrates and phosphates. Having broken through this surface sward, the Puffin brings its sturdy legs and feet into play next, sending debris arcing backwards and sometimes showering inquisitive neighbours. The recipients seldom react to this, taking projectile diggings as a minor inconvenience that goes with the territory of colony housework. The innermost claw on the Puffin's foot differs from the other two in being more curved and set horizontally. As such, it suffers less wear and serves as an extra-sharp tool for gouging soil.

This is not to say that total harmony is always the order of the day. In dense colonies, especially where there is pressure on space (sometimes up to four burrows crowd into each square metre, but one burrow per square metre is more normal), burrows can be fiercely contested with

Below: In this colony in the Outer Hebrides, Scotland, the sward has been compacted by grazing sheep, creating the perfect conditions for Puffins to dig collapse-resistant burrows.

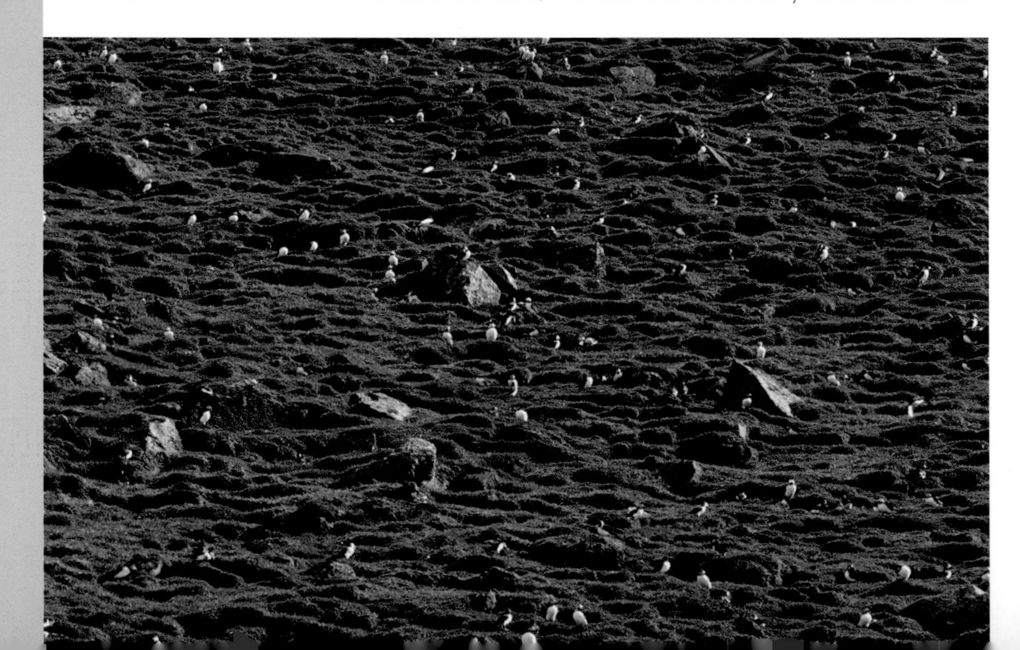

claimants defending their property staunchly against rivals. The defender leans forwards, ruffling its neck feathers to maximise its bulk, flexes its wings and bill-gapes with its tongue raised like a dagger to expose the vivid orange interior, as if to say, 'You'll feel the sharp end of this if you don't retreat'.

Such a threat may be enough for an owner to see off a would-be usurper, but if it is not, turf wars over burrows often escalate into physical fights. The contestants wing-cuff each other and, to the accompaniment of growling calls (a low 'arrr'), grapple with beak and claw. If you ever handle a Puffin and leave your skin exposed within its lunging distance, be prepared for the plier grip and needle sharpness of those latter two weapons. Potentially exhausting if not downright hazardous, fights are typically brief encounters, but not always. A fight is a spectator sport, quickly attracting bystanders. They may gain valuable pecking order information about the prowess of neighbours and even potential mates of the future should they lose or choose to switch their own.

In a serious fight well-matched and equally determined opponents may end up cartwheeling downhill locked in combat. They may sometimes be so preoccupied that they topple over a cliff edge into the sea, until one concedes and the two separate and

Above: Puffins require careful handling to avoid the beak inflicting a painful nip.

Below: The Puffin uses its dagger-like tongue to heighten the aggressive bill-gaping display.

Above: A sequence from a prolonged, no-holds-barred fight on Skomer Island in which the rivals locked beaks and wrestled each other on to their backs.

regain their composure. The protagonists in such fights are often individuals or pairs nearing sexual maturity at around four years of age, for which the burrows of others are highly attractive as ready-made homes. They will claim squatting rights in burrows until they are displaced by their returning owners, whereas younger birds may prospect but are more nervous about such bold trespass. Indeed, two-year-old birds rarely come ashore at all, far less enter burrows.

Because Puffins favour soft, peaty soils, a tunnel needs to be deep enough below the surface not to collapse, although Puffins do sometimes burrow themselves out of house and home (see page 93). Depending on soil type and slope, as well as any extra erosive pressures from trampling by the likes of humans, sheep and Grey Seals, or even inroads by other tunnellers such as Rabbits and Manx Shearwaters, a burrow can persist for several seasons, or may have to be dug, cleared and repaired anew each spring.

In UK colonies the burrow itself is usually a roughly horizontal cul-de-sac up to a metre long, widening at the inner end to form a nest chamber. Especially in the far north of the Puffin's range, where nesting in cliff crevices or among boulders is more common, burrows may be longer and interconnected, forming a labyrinthine maze shared by several occupants. The burrow diameter is typically just sufficient to allow a single Puffin to negotiate at a crouch and defend bodily against unwanted intruders. The tunnel floor often slopes up at the blind end, the better to keep the nest chamber dry (although in torrentially wet summers this may not save it from flooding).

Beyond this there is seldom much embellishment of the chamber. Some Puffin pairs add a thin lining of dried grass, stems, feathers or sometimes flotsam, while others are content to leave the nursery bare. Orange nylon line is commonly part of the nest material – it is mostly collected at sea and ferried back to the burrow. Bits of seaweed and other flotsam are likewise picked up from the water's surface. The most elaborate nests are generally found in rocky substrates, which presumably provides insulation against cold and potential damage to the egg. Puffins do not weave a nest as many land birds do, but a sitting bird will use its beak to crudely rearrange and sort the bedding material.

Above: Floating seaweed contributes to the ill-assorted jumble of debris which passes as a Puffin nest.

Left: A Skomer Puffin scurries to its burrow with a good haul of nest material.

Wooing a mate

Above: Courtship starts on the surface waters around the colony.

There is much more to egg laying, however, than just hewing out and decorating the nursery. Puffins have an elaborate courtship of subtle and expressive body language which, as in Kabuki actors, accentuates the signalling power of their multicoloured facial masks. For a male and female arriving back at the colony with a track record of having bred together in one or more previous years, courtship and copulation generally start in the rafts offshore of the colony. Conducting these preliminaries at sea enables pairs to enter the colony almost ready to lay, which has clear survival value given that there is greater danger to survival on land than at sea. The raft is also the dynamic maritime dance floor where bachelors can start the quest for partners.

In the social whirl of the raft, the male takes the initiative and, having singled out the object of his desire, tries to herd her away from the bobbing throng. Swimming alongside her, he starts jerking his head rapidly skywards, accompanied by 'arr' calls, and flutters his wings in short bursts. Such head flicking is also the come-on signal used on land. The solicited female typically acts coy at first, diving to evade the suitor's advances and resurfacing in the anonymity of the crowd. Unpaired males cruise around in the raft, head-flicking at one female after another in search of any who will return their attentions. If a male's powers of attraction succeed, he mounts the swimming female briefly to copulate. In fact copulation at sea is said to be the norm in Puffins. The male sometimes mounts the female in the colony, but such attempts are not thought likely to result in successful copulation.

On Skomer Island, however, apparently successful copulation on land is not uncommon and even seems to be the norm, especially late in the evening when literally hundreds of birds can sometimes be seen mating. Dave Boyle has collected a remarkable portfolio demonstrating the elaborate ritual of display that serves as a prelude to terrestrial mating on Skomer. Just as in mating at sea, the male, with his chest and sometimes his cheeks puffed out,

Above: On land, the male puts on an elaborate display to invite his mate to copulate.

head-flicks and flutters his heraldically outstretched wings as he approaches the female, never taking his eyes off her for a moment. He may also nibble gently under her beak. If the female welcomes his advances, she may nuzzle back and he will wing-flutter more intensively before she finally crouches and he briefly mounts to copulate.

On water or on land, another striking and common Puffin display, especially when mates greet each other, is 'billing'. It typically starts with a bird of either sex approaching another in a bowed posture and swinging its beak from side to side. It may then nuzzle or nibble the orange rosettes at the base of its partner's beak, a gesture of such intimacy that it could only be tolerated once there was a well-established bond between the two. This is a precursor to both then briskly clashing their beaks broadside against each other for anything from a few seconds to two or three minutes at a time. The sound of

Above: The greeting display between well-established mates is an intimate nuzzling and fencing of beaks.

clashing bills is clearly audible from a distance of several metres and has been aptly likened by Kenny Taylor to the 'clacking of muffled castanets'.

The pioneer seabird biologist Ronald Lockley recorded a male on Skokholm Island billing his mate so vigorously that he pushed her down the slope. A bout of billing often breaks out when a bird lands by its partner at the burrow entrance or the club. It may stimulate nearby birds to start billing as well in a public demonstration of partnership. When Puffins bill on the sea's surface, sometimes as a precursor to copulating, they tend to pirouette, perhaps because they are propelled by the force of hitting their beaks together.

Puffins sometimes approach each other to start billing with a highly ritualised gait called the 'pelican walk', which more than any other display resonates with the Puffin's 'clown' moniker. Standing stiffly erect with head tucked down into its puffed out breast and its tail often cocked, the Puffin pads around with slow, exaggerated steps, raising each foot high as it goes, 'as if', Kenny Taylor puts it, 'the bird was treading on hot coals'. The pelican walk is especially used by a bird alighting at its own burrow entrance (or when a neighbour lands or passes close by), telling others in no uncertain terms that 'this is my property'.

In similar circumstances, the Puffin may employ a variant of the pelican walk in which, rather like a guardsman marking time, it stomps up and down on the spot, one foot after the other, with the webs of its feet conspicuously spread. In both the pelican walk and foot stomping, the Puffin is surely maximising the signal value of its impossibly orange legs and feet, just as the Blue-footed Booby does with its bizarre foot-waving display on tropical islands.

Above: In the 'pelican walk' display, the Puffin goose-steps to advertise its property rights.

Guarding against cuckoldry

In the colony, courtship continues to develop and cement relationships, and to strengthen existing ones. In established pairs, it is also part and parcel of the male sending a 'keep-off' signal to others that his female is spoken for. Even if he drops the intimate language of courtship, however, the male remains highly attentive to his mate, never straying far from her side and watching her like a hawk, so to speak. For Puffins and other colonial seabirds, such 'mate guarding' is crucial in reproductive terms as promiscuity is rife, whether on land or in pre-season rafts at sea. Male Puffins, ever with an eye to the main chance, are alert to the possibility of a sneaky copulation with a fertile female if her mate drops his guard. So, after inseminating his rightful partner, the male Puffin's paramount aim is to avoid the cuckold's fate and ensure that the egg his mate lays carries his genes and not those of some interloping opportunist.

Below: Once paired, the male keeps his mate close to his side to prevent any interloper from fertilising her egg.

From Egg to Puffling

Like 'The Borrowers' of our childhood tales, the subterranean life of the Puffin is a secret netherworld that adds to the mystique the bird holds for us. In the dark cul-de-sac of the nesting burrow, the female Puffin lays a single, pale nondescript egg measured in size by the nourishment she was able to forage from the sea. Both parents share the tending of the egg for the six weeks it takes to hatch into a fluffy Puffin chick, or 'puffling'.

Just as CCTV is laying bare our own everyday and sometimes bizarre urban street behaviour, so the same remote filming technology has been adapted by researchers to shine a lens on the domestic rituals of Puffins underground, a routine for the most part benign but at times darker. The chick is fuelled by energy-rich fish which the parents ferry ashore by the classic Puffin beakful. In just a few weeks the fluffy ball at hatching develops to a state of such self-reliance that on its maiden flight from the burrow it severs all contact with its parents.

Opposite: A young Puffin ready to fledge, some three months after its mother laid her egg.

Below: On hatching, the baby Puffin resembles a furball.

Above: Like the Manx Shearwater and most other burrow-nesters, the Puffin's egg has no need of colour or camouflage.

The preliminaries of burrow occupation over, the female is ready to lay. In UK colonies egg laying generally starts in April, but this varies between colonies. The further north you go, the later the onset of laying, so that in Greenland the first eggs are not laid until early June, the snow and ice cover often not permitting an earlier start. Puffins lay only a single egg, dull white in colour, and initially often with pale brown or lilac markings, although these fade during the course of incubation. In laying a whitish egg, the Puffin is typical of most burrow- and hole-nesting seabirds (such as Manx Shearwaters and Storm Petrels, with which Puffins often co-exist on islands), as well as land birds like owls, woodpeckers and kingfishers.

At least two factors may explain why all these hole nesters have evolved white rather than patterned eggs. Firstly, being hidden in the dark and concealed from aerial predators, they have no need of the camouflage that helps protect eggs of 'open' nesters like gulls and terns. Secondly, the Puffin's nesting habit is different from that of, say, its auk cousin, the Common Guillemot, which not only has striking hieroglyphic markings on its single egg, but these also vary markedly from one egg to another, with no two eggs being alike. This enables the parent Guillemot to distinguish its own egg from those of its neighbours, the eggs being densely packed on a

Right: Puffins lay a single large, nondescript egg.

cliff ledge. Unique egg recognition in the Guillemot has evolved as a safeguard against any pair accidentally incubating another pair's egg and possibly going on to foster the unrelated chick to fledging. With each Puffin pair at home in its bespoke burrow or rock crevice, the risk of such confusion simply does not arise. Unlike the Guillemot, the Puffin has no worries that its energies might be wasted on raising the genes of a total stranger. The Puffin is not capable of recognising its own egg or even, for that matter, its own chick.

The egg laid by a Puffin in UK colonies is roughly 6cm (2⅓in) long by 4cm (1½in) wide at the broad end and weighs around 15 per cent of the adult's body weight. Applying this ratio to the human race would equate to a Goliath baby of 7–8kg (15–18lb). Admittedly this is a crude extrapolation, because weights of individual adult humans and their babies are much more variable than those of Puffins. However, it does serve to underline that egg production is really costly for the female Puffin, who has to dig deep into her body reserves to form an egg capable of hatching into a sturdy chick. To do this, the female ideally needs to arrive back at the colony in spring in good body condition and find plenty of fish to eat locally to cope with the additional nutritional burden of pregnancy.

The size of the Puffin's egg is therefore a litmus test of the quality of the marine environment and its food supply. Environmental change and its implications for Puffins is explored later (see pages 90–107), but evidence that all is not well has emerged from the worrying finding by Rob Barrett and his fellow scientists at Norwegian colonies (mirrored on the Isle of May in Scotland) that Puffin eggs have been getting progressively smaller over the last 30 years. Young Puffins tend to nest later and lay smaller eggs than older, more experienced birds, but this does not explain the long-term decline in egg size.

Once the egg is laid, both sexes share the task of incubation, each pair working out its own shift pattern, with some individuals tending to be tighter sitters than others. At one Norwegian colony shifts varied from 23 hours to a marathon 80 hours. However, the sitting bird is not saddled with the egg all that time, and takes a periodic

Above: Unlike the Puffin's, the Guillemot's egg is colourful and uniquely patterned for instant recognition on a crowded ledge.

Below: When first laid, the Puffin's egg is at best lightly blotched.

Above: If there's a splattering of droppings around the entrance, it is more than likely a burrow is occupied.

toilet break at the tunnel entrance, leaving telltale white streaks showing that the burrow is occupied. Puffins on incubation duty may even take the liberty of a short foray offshore, perhaps to bathe and preen, before resuming their stint underground. In fair weather the burrow is such a good cocoon that the egg does not suffer from being left unguarded for a limited period. The off-duty bird typically roosts at sea or sometimes at the edge of the colony.

Partners tend to swop incubation shifts in the morning or evening. Up until a few years ago, details of the changeover and other burrow behaviour remained largely a mystery, but our understanding has been transformed by the 'Puffin cam'. This is an endoscope similar to that used in hospitals, used to literally shine a light into the nest chamber. A miniature digital camera lens lit by infrared (a spectrum not visible to birds) is mounted at the end of a long, flexible tube, which is carefully inserted into the depths of the burrow. The camera then transmits images of the innermost contents and happenings to an outside monitor, which the field worker can then examine. Puffins readily tolerate an endoscope inspection, whereas manhandling them out of their burrows is much more stressful and can even lead to nest desertion.

Above: A Puffin about to leave the burrow to allow its mate a stint at incubating.

The endoscope is portable from one burrow to the next, but a camera can also be inserted in the burrow as a fixture, usually housed in a box for protection, for the duration of the breeding effort from nest building to fledging. The small CCTV device does not interfere with the burrow occupants, who carry on as normal. By then linking the camera to a computer, live footage can be beamed to the outside world, giving researchers and the public an unprecedented window into the Puffin's private subterranean life. Such 'webcams' are now routinely installed in selected burrows at the RSPB's seabird colony reserves at Sumburgh Head (Shetland) and Coquet Island (Northumberland), as well as by those responsible for managing numerous other Puffin colonies on both sides of the Atlantic. On the Farne Islands the National Trust has a webcam ingeniously housed in the tummy of a replica Puffin, the better to get up close and personal to the real thing.

While webcams can reveal unsuspected dramas underground, mostly they enable us to appreciate as never before the everyday intimacies of family life, such as incubation changeover. Now we can see the sitting bird caress its beak against that of its returning partner,

Below: While Annette Fayet inserts an endoscope into a Skomer Island nest chamber, her field assistant examines the transmitted image.

Reality TV down under

Above: Burrow occupancy revealed by an endoscope on Coquet Island is used to census the colony.

Every day the live Puffin burrow webcam at the RSPB's Sumburgh Head colony is clicked on by thousands of avid followers for whom it has become compulsive viewing. As warden Helen Moncrieff said: 'I have to offer a word of warning. Watching a sleeping, sighing, shuffling Puffin can become addictive but is not harmful to your health.'

In 2012 the live hatching of a chick on the webcam attracted a worldwide audience of 5,000 viewers from as far away as Australia and Chile, some of whom remained glued to their screens for hours at a time. However, the public also witnessed the uglier side of burrow life, and the struggle Puffins have to raise young, especially in an era of tougher environmental conditions. A month later the webcam showed the chick dead and the parents visibly distressed, kicking their nest material around. Although the puffling had appeared healthy up until its demise (it was a good weight for its age), it had telltale wounds on its back, indicating that it may have been attacked by an interloper.

In 2010 there had been a foretaste of this when a fully grown adult, nicknamed 'Asbo Puffin' by the warden, was seen on film repeatedly gatecrashing a burrow to bully the resident chick. Despite being subjected to kicking and pecking, the youngster luckily survived to fledge later on. In summer 2013 the webcam again failed to chart a successful breeding attempt, revealing a badly cracked egg before the parents pushed it out of the burrow. It appears that an adult was responsible for damaging the egg, but it is not clear if this was the result of an accident or something more sinister.

Above: Using its beak with great delicacy, a Coquet Island Puffin manoeuvres the egg under its body.

Left: A webcam will reveal intimate moments like a partner arriving to add nest material to the nest chamber.

who arrives bearing gifts – a stem or two of dried grass to add to the nest. The sitting bird then eases itself off the egg and exits the burrow, allowing its partner to take charge of the precious cargo. At this point we traditionally think of birds as tucking the egg (or eggs in species which have a true clutch) under the warmth of their breast plumage, but the Puffin has a surprise. Using its beak, the bird instead gently nuzzles the egg into the 'armpit' of one of its folded wings to nestle against a special patch of bare skin. This 'brood patch' is richly supplied with blood vessels that transmit body heat to the egg. Through localised feather loss, the adult Puffin develops two such patches, one under each wing, shortly before egg laying. The feathers grow back again after hatching, when the chick is old enough not to need brooding.

The question arises as to why our Atlantic Puffin, in common with the other three puffin species, has two brood patches when it lays only one egg. Perhaps historically the proto-puffin tribe did have a regular clutch of two, but over time evolved to invest in one bumper egg while retaining two brood patches as an anatomical echo of the past. While modern Puffins never lay two eggs at one sitting, we do know that if the single egg is destroyed or taken soon after laying, some Puffins will lay a replacement a couple of weeks later. However, if the egg is lost far into incubation, there is no possibility of laying another and for that pair the breeding season is a write-off.

Tunnel of love

Above: The white egg tooth at the tip of the chick's upper mandible helps it chip its way out of the egg.

The Puffin's incubation is a lengthy affair lasting about six weeks. In the last few days the pair can anticipate the next crucial phase of parenthood as the chick begins to chip its way out of the egg. The beak of the baby Puffin – or 'puffling' – is not hard or strong enough to do this on its own, so it pierces the eggshell with the aid of a whitish, specially hardened and sharp cutting tool at the tip of its upper mandible. This so-called 'egg tooth' may survive for several weeks before it is lost. The chick does not emerge naked (as do the chicks of many land birds and even some seabirds), but as a cute, fluffy ball of long, soft down, greyish-black in colour except for a white belly patch. The hatchling's beak is a stubby black cone with no hint of the riot of colour that will adorn it in future, but even at this stage the crinkled, fleshy rosette at each corner of the beak is evident.

On hatching the chick retains a reservoir of food from the egg's yolk sac as a buffer in the crucial first hours of its life. Once this literal nest egg of food is spent, however, the chick, with its dark eyes now fully open, begins to pester its brooding parent for food with plaintive peeping calls. The parents now switch into food-provisioning mode, although they will continue to ensure that their offspring is brooded more or less continuously for another week or so until it is capable of maintaining its own body temperature unattended.

Brooding during these vulnerable early days is also the best safeguard against any hostile intrusion into the burrow. Up until near fledging, the inner sanctum of the nest chamber remains the chick's safe haven, to which it scuttles back if brought to the burrow entrance. With predatory gulls on the lookout for a tasty snack, hazards abound at that portal to the outside world, so a small, downy Puffin chick never strays voluntarily to the burrow entrance unless driven by hunger, when its desperate begging calls may prove to be a fatal attraction to predators.

This enforced incarceration, however, raises personal hygiene issues, and to avoid soiling itself and its nursery chamber, the chick habitually uses an en suite latrine area situated in a recess of the burrow. From decades of hands-on experience, Mike Harris and Sarah Wanless reflect wrily in their classic Puffin monograph on the occupational hazard this poses to the field worker: 'the latrine makes the job of extracting a Puffin chick from a burrow particularly messy and smelly since it invariably seems to be situated just where he/she needs to rest their

Above: The juvenile plumage grows until only a few vestigial wisps of fluff remain.

Below: With the ever present danger of gulls, small Puffin chicks rarely venture from the nest chamber to the burrow entrance.

Above: The puffling's en-suite latrine ensures that most of the burrow remains unsoiled.

elbow'. As the chick gets older, it may abandon its foetid toilet for a spot closer to the burrow entrance where conditions are more hygienic. This is more than just a nicety – any loss of waterproofing caused by plumage soiling as the bird gets ready to fledge could prove fatal when the bird first enters the sea.

When it hatches the Puffin chick weighs around 40g (1⅓oz), or about the same as three £2 coins. By the time it fledges some six weeks later, its body weight will have increased as much as eight-fold to around 320g (11¼oz). However, just how long it takes to finally quit its burrow, and how heavy it is at that point, varies a lot depending on a host of factors – most importantly how well the chick has been fed from hatching onwards and whether it was raised earlier or later in the year.

To achieve this remarkable increase in body weight over just a few weeks, the chick's parents fuel it with energy-rich oily fish, which is ferried from the sea to the burrow by the beakful. A Puffin with a serried row of glistening, silvery fish drooping bootlace-like from either side of its parrot beak is surely one of the most iconic and instantly recognisable images, not only of a Puffin but also of the bird kingdom at large. In the next chapter there is more about how the Puffin achieves this feat of food finding and storage, and which fish species are involved.

Below: The parents are a conveyor belt of oil-rich fish for their offspring.

Above: The adults adjust the size of fish to the age and size of their chick to ensure ease of handling.

In the first few days of the chick's life, the parent brings in small, made-to-measure fish that are easy for the newborn to swallow. At each sitting older chicks are delivered fewer but much bigger fish, providing they are available in the surrounding waters for the parents to catch. Flying in from the sea, and especially when it is carrying big, eye-catching fish, the parent may circle its burrow area a few times before seizing an opportune moment to run the gauntlet of any marauding gulls and make a beeline for its burrow.

With the certain knowledge that it is 'home', the Puffin immediately scuttles underground to safety, sometimes repeating a special lure call that sounds like a soft clicking. The arrival of food is met with great excitement by the chick, which ramps up its own 'peep' calls, as if to remind its parent how needy it is. Very young chicks start ingesting from the front of the proffering beak, the precious fish passing deftly from bill to bill. Chicks often sleep after a meal.

As the chick ages the first few fish may be transferred like this, but the parent then typically drops the rest unceremoniously on the burrow floor for the chick to

Above: A Puffin lowering its 'undercarriage' in preparation for landing with a meal of fish.

retrieve on its own. Once the youngster is big enough to defend itself against cold-calling gulls, its parents increasingly offload food like this just inside the burrow entrance. Self-service from the ground must assist the puffling to develop valuable manipulative skills. Likewise, from an early age, chicks pick up and toss around scraps of nest material, and this presumably gives them an early taste of what it might eventually be like to grasp that first slippery sandeel.

The chick's daily intake is dependent, of course, not just on fish size and meal size, but also on how many meals the parents can deliver in a day. On Skomer Island, Ruth Ashcroft discovered that in the first few days after hatching, when a chick's needs were modest, it was fed only two to three times a day, but as it developed and its demands increased, the parents responded accordingly, upping their visitation rate to about 10 feeds per day – but this does not escalate until fledging. Intriguingly, at around four weeks old, a week or more before fledging, most chicks reach their peak weight in the nest and thereafter slim down a bit, losing up to 10 per cent of their body weight. This is a commonly observed pattern in other auk species and indeed in several other families of seabird. It may be the bird equivalent of losing puppy fat to hone the body for the rigours of daily survival in the outside oceanic world.

In the pioneering days of Puffin research, the weight loss was thought to arise from the parents switching off the food supply and deserting the chick in the run-up to fledging. This was certainly the belief of island adventurer Ronald Lockley, whose pioneering fieldwork on Skokholm Island transformed our knowledge of Puffins, Manx Shearwaters and other seabirds, blazing the trail for future researchers. However, his interpretation that the parents stop feeding their chick, which then 'remains fasting in the burrow for several days', turns out to have been mistaken. We now know that the parents keep provisioning their youngster every day until it fledges.

By the time the Puffin chick is ready to fledge, it looks just like a slighter model of the adult except for being a monochrome version and lacking the adult's rainbow colours. Its plumage is now fully equipped for flight and immersion in the sea, sporting at most a random tuft of fluff left over from its puffling days. In the last few evenings before finally abandoning the cradle of its burrow (and earlier if for some reason it is underfed), the chick increasingly ventures to the entrance to take the air, acquaint itself with the fresh perspectives that daylight brings and experience the soap opera of colony life.

Below: The fledgling is a slighter and duller prototype of its parent.

These forays to the surface may also enable the chick to configure the dynamics of the sun and stars for future reference when direction finding is called for on the high seas, and possibly to get a fix on the colony for returning in seasons to come. Unfettered by the confines of its burrow, the new freedom of stepping just outside the burrow also gives the chick full freedom to exercise its wings. A parent typically sits nearby in a chaperone role, communicating with low-key calls, and ever ready to raise the alarm if danger beckons, sending its offspring scurrying back down the burrow.

If the chick should stray into a neighbouring territory, the resident adult reacts aggressively and sends it packing. That said, one of the more remarkable facts about Puffins, differentiating them from most other seabirds, is that adults do not recognise their own chicks. In the majority of seabird species, recognition is based on voice. Each chick's call is a unique signature tune to its parents for distinguishing and unerringly locating their own offspring in the noisy maelstrom of a colony or in wide open spaces. Indeed, in the Puffin's auk relative, the Common Guillemot, adult chick recognition starts while the chick is still in the egg, calling audibly through the shell and, in turn, fixating on the voice of its parents. This is a vital bonding mechanism given the dense packing of Common Guillemots nesting on cliff ledges, and the consequently

Below: As fledging nears, a Bempton Cliffs (Yorkshire) puffling spends time at the burrow entrance, chaperoned by a parent.

Above: An adult rebuffs a neighbour's chick that has strayed into its territory.

high risk of an adult expending its precious parental care on the wrong chick.

Mutual voice recognition in the Guillemot and many other seabirds also serves the extended period of post-fledging care, enabling parent and offspring to find each other after they leave the colony. Herein lies the probable explanation for why adult Puffins fail to recognise their own offspring, because adult Atlantic Puffins, in keeping with the other three puffin species, do not care for their young after they fledge. In other words, young Puffins leave the burrow on their own and from that moment on are entirely independent of their parents.

Fledging happens in July or August, depending on the latitude of the colony, and under cover of darkness to avoid the attentions of gulls, skuas and other would-be predators. The fledgling Puffin is a fully functional bird, ready to take to the wing on quitting the burrow, and making its maiden flight from the colony slopes out to the relative safety of the sea where, if all goes well, it splashes down some distance from shore. The higher above sea-level the elevation of its natal colony, the further its diagonal trajectory is likely to take it offshore. Once in the water juveniles are immediately able to swim, dive and fend for themselves. Given the weeks and sometimes even months that many other seabird species devote to

Above: Pufflings usually fledge at night but this one made a bolt for freedom in broad daylight.

Below: Alighting on the sea after its maiden flight, a puffling flexes its wings for the onward journey.

their offspring after fledging to show them the ropes of where and how to catch fish efficiently, it is remarkable that the young Puffin leaves the cradle of its burrow fully equipped to deep dive and chase down a mercurial fish it has up till then only seen dangling lifeless from its parent's beak. It is an outstanding example of nature trumping nurture to programme the next generation.

The pufflings' parents may fail to switch off their feeding drive immediately, however, and not uncommonly return to the burrow in the morning with a meal of fish, only to find that their charge has left home. Most adults desert the colony and head offshore not long after their young, although some continue to visit their burrow for several weeks. By mid-August UK Puffin colonies are largely silent and less colourful again, except for a few late nesters still nurturing their young in a race against time to beat beckoning autumn gales. There is no evidence that dispersing pair members do anything other than go their separate ways on the high seas for the next seven or eight months, until the breeding circus returns and starts all over again.

Puffin patrol

On St Kilda I was privileged to take part in a nightly mercy mission during peak fledging time for Puffins. On uninhabited islands, young Puffins fledge from their burrows in darkness (or at least in what Scots call the 'simmer dim' of short northern summer nights), so there are no distractions from making a beeline to the sea. On St Kilda, however, the army base had an electricity-generating station that emitted low-frequency sound and was lit like a beacon throughout the night. This must have disorientated Puffins fledging from the surrounding slopes for they were attracted like moths to the base, especially on murky nights when visibility was poor. By midnight we could routinely anticipate the surreal spectacle of numerous fledglings tottering around the buildings like street urchins, with no chance of making it to sea under their own steam. Our job was to collect these waifs and strays in cardboard boxes and keep them safe indoors until dawn. On nights when there was an exceptional invasion, we sometimes ran out of boxes and popped some Puffins in the bath.

At daybreak, when there were no longer any lights to divert the birds, we would trek to a high cliff-top with our precious boxed-up cargo and, one by one, launch the fledglings seawards on the flight they should have made the night before, had everything gone to nature's plan. With the advantage of the cliff's height, they flew like arrows to the sea at a pace likely to outstrip any piratical gulls, yet we always followed their paths with binoculars to make sure they landed safe and sound on the water. Over the course of the summer, we rescued hundreds of young Puffins.

A similar Puffin patrol has been an annual ritual for generations on Heimaey, the largest of Iceland's Westmann Islands. After mid-August the children mount a search-and-rescue operation for pufflings (called *pyjsa*) lured into town at night by its lights. Roving the streets in family parties, they store the orphans in boxes and, come daytime, release them by the hundred from the seashore. Heimaey's children have all been taught how best to launch the Puffins to freedom, as if throwing paper darts. Historically, vested interests may have encouraged this conservation action, because Puffins are also harvested on the Westmann Islands (see page 102).

Much closer to our shores, pufflings from Craigleith Island in the Firth of Forth fetch up every year in streets and gardens in the Scottish coastal town of North Berwick. In August 2013 one was even found wandering the corridors of a hotel. The Scottish Seabird Centre based in the town appeals to locals to search under their cars – a favourite refuge – and report any strays to their 'Puffling Rescue Service'. In 2013 a handful of birds were taken into care and ferried offshore to be returned to the sea in the early evening, the time of day when the Centre has found that releases are most likely to succeed.

Above: Pufflings collected from the army camp on Hirta, St Kilda, are billeted overnight before early morning release.

Flying and Foraging

Puffins are essentially sea-going creatures obliged to make an annual landfall to breed. Not only are they formidable globetrotters over the ocean's surface, but they also plumb its depths to hunt for fish. The Puffin has therefore evolved a body design that is effectively a compromise between these conflicting challenges and demands. Its wings have to serve as both successful aerofoils for flight and paddles for diving, and must be able to withstand a punishing work rate in both these mediums.

This chapter discusses these trade-offs, and the Puffin's remarkable adaptations and prowess for deep diving, and for catching and transporting fish. The waters around our coast offer a varied menu of marine prey, but Puffins are highly dependent on a very narrow fish spectrum, which they harvest with great efficiency. Among seabirds the Puffin truly ranks as one of our most accomplished deep-sea trawlers.

Opposite: The Puffin is a highly efficient fishing machine.

Below: Puffins are essentially mariners who traverse the boundaries of sea, air and land in the breeding season.

Above: Wings must work hard to generate the lift needed for getting a stocky body airborne.

As noted in Chapter 'Vital Statistics', Puffins are not the most manoeuvrable of fliers, compared with, for example, terns and skuas. They are not alone in this as they share a body design that is fundamental to all the auk family. To what do we attribute this lack of aerial dexterity? In their treatise *The Auks* (1998), Tony Gaston and Ian Jones put it eloquently when they say:

> More than any other birds, they [auks] are a product of both sea and air. The only other birds that have travelled so far in adapting to the marine environment, the penguins, have forsaken flight completely. The auks now . . . teeter on the brink between flying and flightlessness, squeezing the last ounce of underwater performance commensurate with retaining the ability to fly.

Below: Especially when there's no headwind, Puffins use the sea surface as a runway to get aloft.

As a result of the combination of the Puffin's stout body and its stubby wings, resulting in what is technically called a heavy 'wing loading', Puffins need a fast, whirring wingbeat to generate enough lift and air speed. A Puffin can beat its wings at up to 400 times per minute, generating a forward speed of up to 80km/h (50mph). The Puffin's flight style is thus aptly described as 'beetling', as if the bird is a clockwork toy hurtling along as fast as it can just to stay airborne. The inertia a Puffin has to overcome to get airborne is especially evident from its foot-pattering

take-off from the sea's surface in calm conditions when there is no wind-assisted lift.

When a Puffin slows down to alight on land, it has to muster every scrap of lift at its disposal to avoid stalling. Here it makes another use of its webbed feet, spreading them out in line with its body to yield a little extra lift. Equally, lowering the feet serves as an air brake when necessary, and as the bird lowers its undercarriage for landing, its whole body swings from horizontal to near vertical to slow its forward momentum. Puffins positively seem to enjoy flying when they can lean into a stiff breeze, which generates lift without the need for manic wing flapping. On breezy days in the colony, Puffins dance aloft like marionettes, an idyllic scene for any Puffin watcher.

Above: To slow momentum for landing, the wings and body assume a vertical crucifix posture.

Below: At slow airspeeds and in updraughts, the outspread feet generate a little extra lift and rudder control.

Above: In its 'moth flight' display, the Puffin demonstrates a grace that belies its solid frame.

Unlike in terns with their sinuous wings, aerial display does not feature much in the Puffin's behaviour repertoire, but it does have one distinctive flight ritual – the so-called 'moth flight'. Moth flying may be used by a bird of either sex as it takes off from the colony or from a 'raft' of Puffins at sea. The Puffin flies with its wings angled stiffly upwards, its wingtips fluttering rapidly in a shallow arc. With this minimal wing action, normal flight speed is about halved. A moth-flying Puffin's body is arched with its head angled down and, oddest of all, its feet are typically crossed.

Given the Puffin's heavy wing loading, moth flight must be energetically demanding and is not sustained for long, but it cannot be mistaken for normal flight so it clearly sends an important message to other Puffins. Moth flight is seen most often over the colony slopes at dusk, and appears to be a 'come and join me' signal from a bird intending to head seawards to roost for the night. If its first moth flight fails to attract others, a displaying bird will circle the colony area repeatedly until it succeeds in forming a posse moth-flying in unison, whereupon all the birds head offshore together. Moth flight may have evolved to facilitate communal roosting, which clearly confers some benefit for Puffins.

For a seabird that is not the most composed in the air, the Puffin's moth flight is a thing of balletic grace, never better captured than by Kenny Taylor's description of a magical scene at a large Hebridean colony early in the breeding season:

> It was a calm night, with a peachy afterglow behind the mountains to the west. Thousands of puffins were in the wheel above where I stood, silhouetted against the pale sky. Looking closer, I realized that many of these birds were in pairs, moth flying, almost touching each other in the still air. The sense of peace conveyed by this movement – massed but carefully balanced and tranquil – was overwhelming.

Diving for dinner

Seabirds have devised various ways of exploiting shoals of small fish. Lightweight seabirds like terns and kittiwakes are surface feeders, at best plunge diving from the air to a depth of a metre (three feet) or so to seize fish or crustaceans in their beaks. Gannets are the most spectacular plunge divers, plummeting headlong from the air, often from considerable heights, and after they make splashdown they are also capable of propelling themselves underwater to a depth of around 20m (66ft).

Unlike these birds, the auks including the Puffin are not designed for momentum diving from the air, but are highly accomplished submersible pursuit divers, making their entry by 'duck diving' from the surface, then chasing down their prey under the sea. Using its wings to propel itself, and its webbed feet at the stern for fine rudder control, the somewhat ungainly brick flying in the air is transformed into a water sprite.

To witness a Puffin pursuit diving is to be immediately struck by the realisation that it is effectively flying underwater, and that its wings are designed as much to be oars as they are to be aerofoils. In essence, the Puffin is, if anything, more at home in water than it is in the air. In medieval times there was a belief that a Puffin was actually a cross between a bird and a fish because of its underwater

Above: Whereas the Gannet plunges headlong, the Puffin duck-dives from the sea surface.

Below: Off the Farne Islands, the Puffin's wings demonstrate their dual function, beating a path to fish shoals.

mastery. This allowed some people to eat Puffins at Lent and on Fridays, because they could avoid the Catholic Church's prohibition on partaking of meat at these times.

A Norwegian study found that, underwater, Puffins could swim horizontally at 1.5m (5ft) per second or about 5km/h (3mph). Humans average 1–2km/h (2/3 –1¼mph), at least when swimming on the surface, so a Puffin beats us hands down. Nor can we hope to match the depths a Puffin can attain, at least not without an oxygen tank and other artificial aids. For decades our only knowledge of how deep Puffins could dive came from birds caught in fishing nets set at known depths, but with the development of radio-tracking and depth-recording devices in the 1980s, we now know that Puffins can make dives to depths of 50m (164ft) and exceptionally to 60–70m (197–230ft), lasting over five minutes, although most of their hunting is done at depths much shallower than this (see box, page 82).

Even so, given the water pressure at such depths, and the challenge of keeping the muscles oxygenated for prolonged periods of immersion, this is a remarkable feat for a relatively small bird. How does it do it? In common with their close relatives, Puffins have evolved

Below: Dense plumage insulates the Puffin for deep diving.

a number of physical and physiological adaptations for deep diving. The flight constraints imposed by the Puffin's relatively heavy body for its size have already been noted. This characteristic is attributable to dense, heavy bones which, like in a scuba diver equipped with a weight belt, counteract the natural forces of buoyancy and help the bird descend quickly into the depths.

Puffins also have unusually thick plumage (it has been likened to a pile carpet), which not only streamlines them but also serves as an insulating duvet against chill sea temperatures. Last but not least, in common with seals and whales, Puffin muscle is rich in myoglobin, a super oxygen store that enables the bird to make extended deep dives, and is capable of being recharged within seconds at the sea's surface for the next dive. Puffins and other auks tend to make a series of dives one after another, followed by a prolonged rest to recuperate.

A typical foraging trip for a Puffin in the breeding season starts with a group embarking from the colony for an area offshore, probably a destination where one or more of the flock has had success in finding sufficient food of the right kind before. As electronic tracking devices for birds get progressively smaller and lighter,

Below: A posse sets off on a hunting trip.

Speed-diving athletes

Mike Harris and Sarah Wanless have pioneered the use of electronic technology to reveal the remarkable diving performance of Puffins breeding on the Isle of May off the Fife coast of Scotland. Using VHF telemetry, they were able to discover where the birds went, and also – because VHF signals do not travel through water – the gaps in transmission of 'bleeps' when they were submerged. One bird made a record 194 dives in just 84 minutes. Over a longer series, however, the average dive time panned out at 28 seconds, with just six seconds on the surface between each dive to inhale before the next plunge.

The Isle of May researchers have also attached miniature devices called time-depth recorders (TDRs) to Puffin leg rings to unlock more of the Puffin's diving secrets during chick rearing. Once retrieved, the TDRs showed that over the course of a day Puffins typically make large numbers of very short dives, mostly to just 4m (13ft), with few exceeding 7m (23ft) and the maximum being 20m (66ft).

Innovative technology in the shape of miniature cameras is additionally shedding light on the diving feats of seabirds. In 2012 a camera attached to the back of a South American Imperial Cormorant showed it diving to 46m (151ft) in 40 seconds, catching a fish and returning to the surface, the round trip taking 160 seconds. The researchers proclaimed that if speed diving were an Olympic sport, the Imperial Cormorant would surely win a medal! However, if the cormorant won the sprint, the marathon would be won every time by the Emperor Penguin, which has been recorded diving to 550m (1,805ft) and holding its breath for up to 22 minutes. To achieve this, the Emperor Penguin capitalises on the same adaptations that equip the Puffin for diving.

In addition, the penguin's heart rate drops dramatically when it dives so that it can eke the maximum out of the oxygen stored in its blood and muscles. Perhaps Puffins do the same.

researchers have been able to chart Puffins seeking out these hotspots with a precision that was only dreamed of by early seabird biologists. Chick rearing is exhausting, so Puffins try to husband their energy expenditure by feeding as close to the colony as they can. Skomer Island Puffins mostly feed within 15km (9 miles) of the colony and none do so more than 35km (21¾ miles) away.

Other studies have yielded similar results, although Puffins from the Norwegian island of Røst radiated as far as 137km (85 miles) from the colony. Mike Harris and Sarah Wanless cite a remarkable at-sea observation in the far north, when Mark Tasker reported fish-carrying Puffins 'flying purposefully' to the Faroe Islands no less than 250km (155 miles) away. Resorting to such extreme distances is a desperate last resort when the food supply has failed within reasonable striking distance of the colony. It is a clear sign of a colony under stress and comes at a cost, as testified by the poor breeding success of Faroese Puffins in the year concerned. We return to the threat of food shortage in the next chapter.

Once offshore, Puffins heading out to hunt will ascertain whether they are on the right track by spotting others beating a path back to the colony with successful hauls of fish in their beaks, or by observing other Puffins actively diving. Once alighted on the surface of the water, a Puffin will often immerse its head and peer down for

Left: Puffins will peer into the water column in efforts to detect fish before they submerge.

Below: Meals for chicks on the Isle of May are monitored by briefly and harmlessly catching a sample of Puffins in fine 'mist-nets' and retrieving what they drop. The fish are wrapped in foil for later identification, weighing and measurement.

signs of life, just as we do when snorkelling. Sometimes, however, it spots from afar a feeding frenzy characterised by a highly animated mix of Gannets, assorted auks, gulls and maybe even cetaceans pummelling a fish shoal near the surface. Faced with such a bonanza, Puffins have been seen crash diving onto the surface and, almost in the same motion, diving without any further niceties.

As visual hunters Puffins are daytime divers, and there is no evidence to suggest that they dive under cover of darkness. Nocturnal feeding is mostly recorded among tropical seabirds because many of their prey species are luminescent. While Puffins forage at any time of day, they tend to hunt most intensively at dawn and dusk. The same diurnal pattern has been found in most of our other seabirds and may be motivated by the hunger of adults and chicks first thing in the morning, and the need for a final feeding boost before nightfall. The Puffin's staple diet of sandeels is also only accessible and on the menu during the day (see box, page 87) so night-time would offer leaner pickings for Puffins, even if their visual powers enabled them to hunt then.

From what they transport back to chicks, we know a lot about what Puffins feed on in the breeding season, but we know very little about their winter diet, which is consumed when they are scattered to the four winds far offshore. The recorded prey spectrum of the Puffin across

Below: Sandeels are the staple diet of British Puffins.

its geographical range is wide, varying from small fish and squid to crustaceans and marine worms, with shrimps and other crustaceans featuring more in Arctic regions. Overall, adults are less selective about what they eat themselves than what they deliver to their young.

Stomach contents of Puffins found dead on the North Sea coast in winter thus included a number of species which chicks would find difficult or downright hazardous to swallow, because they were either spiny (sticklebacks), or very bony and rigid (Snake Pipefish). The latter, which underwent a population explosion in the North Atlantic in 2005–2006, have even been known to choke seabird chicks; even if they can be swallowed, they scarcely offer any more nutritional value than a leather strap.

Chicks therefore need a plentiful supply of easily digestible, energy-rich fish in order to develop and thrive. The two creatures of choice that best fulfil this requirement in UK waters are sandeels (or 'sand lance', as they are called in North America) and sprats, both of which make a conveniently packed, oil-rich ready meal for a growing chick. Weight for weight, sprats and other 'clupeid' fish like herring are the highest energy species on offer, with sandeels not far behind. Chick-rearing Puffins also avail themselves of less nutritious, so-called 'gadoid' fish such as whiting and rockling.

Sandeels have the added advantage of being the ideal

Below: Of two fish brought ashore by a Puffin on the Inner Hebrides, one is an indigestible Snake Pipefish which has wrapped itself around the bird's neck.

size and shape for chicks, being long, thin and easy to swallow. Most prey fed to young in UK colonies varies between 5 and 8cm (2 and 3in) long, but 10cm (4in) items are not uncommon. A massive 20.7cm (8in) sandeel weighing 25g (1oz), the biggest ever recorded in a colony, must have seriously stretched the handling powers of both parent and chick, and surely would have been rejected if destined for a younger, smaller-billed chick.

The balance of the diet varies between colonies, between years and even within the span of a breeding season, but in UK studies at times 80 per cent or more of the fish ferried back to young were sandeels. While Puffins appear to prefer sandeels and sprats for all the reasons mentioned, it is hard to judge the extent to which they actively select them to eat, or whether they just take what is most readily available at the time. Recent research on the Dogger Bank shows that sandeels are highly localised and concentrated on their chosen sandbanks, so for a Puffin it is definitely a case of 'find the right habitat and you're likely to find the right prey'. From past experience, Puffins will certainly know the locations of such foraging hotspots in the hinterland of their colony, and will make a beeline for them as long as they continue to produce the goods.

Anyone who spends time in a Puffin colony during the height of chick rearing will inevitably ponder the 64,000-dollar question: 'Just how does a Puffin bring back all those fish in its beak?' Apart from Puffins, only some tern species occasionally pull off the feat of returning to the colony with several fish in their beak – one Roseate Tern on Great Gull Island (New York) managed an eye-watering nine small fish at once. Whether terns usually catch these fish simultaneously or sequentially is still unclear, and only once in three years of intensively watching terns plunge diving did I see an individual, a Sandwich Tern, dive with a fish already in its bill. Certainly, terns have no special tricks up their beaks for hanging on to one fish after another.

By contrast, the Puffin's beak is highly adapted for keeping a 'velcro' grip on items caught successively underwater. Backwards-facing spines on the roof of the

The retiring sandeel

As their name suggests, sandeels live in submerged sandbanks. These banks are a remnant of the last ice age which surrounded the UK coast. Indeed, they spend most of their time buried in the sand in a dormant state. Apart from a brief adult emergence in winter to spawn, it is only in spring and summer, triggered by rising sea temperatures, that the bulk of the sandeel population wriggles out of the sand into the water column to swim and feed on zooplankton, though even by June or July adults have accumulated enough fat reserves to retreat back into the sand again.

During the summer months sandeels have a daily rhythm in which they are active only in daylight, ascending into the water column at dawn, and descending again at dusk to once more seek refuge in the sand. They undoubtedly do this to conserve energy and avoid predators like the Puffin. They will also bury themselves by day if danger threatens. Once, when I was watching terns catching sandeels in Druridge Bay, Northumberland, the spring tide receded rapidly to reveal an extraordinary sight – a great expanse studded with thousands of tiny, metallic-looking sandeel heads poking out of the sand, as if the beach had suddenly been converted for braille.

So the sandeel's life plan is simple: 'I will only swim free as long as it takes to eat my fill and then I'll hide in the sand from the rigours of the world, including for an extended winter fast'. Sandeels, then, are only available to Puffins during the day, and also only in summer unless they dig them out of the sand in winter (one can imagine the Puffin maybe using its beak like a hoe to plough the sandy seabed!). There is no evidence, however, that they do.

Above: Adult sandeels spend most of their lives buried in the seabed.

mouth and on the tongue lock one fish in place as the hunter goes for another. The lower mandible also has a flexible hinge where it meets the skull, which helps wedge already-caught items against the palate when the beak opens to seize another one. You can imagine the Puffin charging through a shoal of sandeels or sprats (which form a tight ball when attacked), taking one victim after another.

A myth grew up in the early literature that a Puffin's fish load is typically stacked in a symmetrical pattern, with the fish facing in alternate directions (head, tail, head and so on) on each side of the bill. This led to further speculation that the Puffin, in the thick of a shoal of fish all fleeing before it in the same direction, snaps them up by turning its head alternately left and right.

Below: Spines on the roof of the mouth pinion fish as they are caught, allowing the Puffin to pursue the next one.

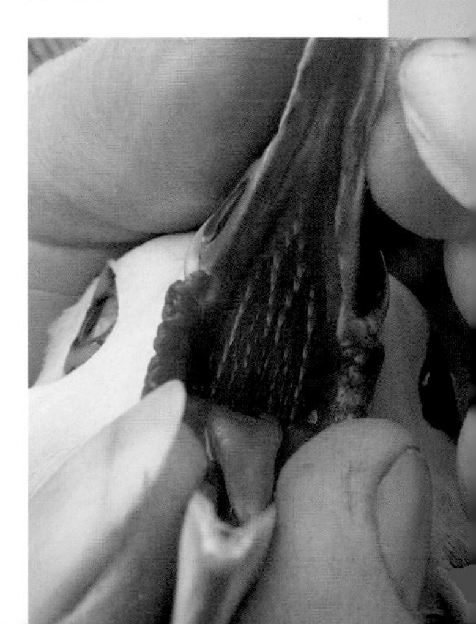

Above: Filamentous sandeels are ideal for mass stacking in the beak.

As Mike Harris and Sarah Wanless attest from having witnessed thousands of meals brought to the colony: 'This nice story is, unfortunately, quite untrue and the fish are arranged haphazardly'.

The record for a single beakful of fish is held by a Norwegian Puffin that brought in a meal consisting of 80 tiny specimens of a herring-like fish called the capelin. When I was studying Puffins on St Kilda, the most I ever recorded was a cargo of 62 tiddlers, all but one of which were filamentous, translucent sandeels, each a mere 4cm (1½in) long. This Puffin's beak seemed to be fringed with splinters of glass, and indeed sandeels are aptly called 'needles' at this tender age. Added together, all these tiny fish tipped the scales at less than 5g (1/5 oz). Even if destined for a very small chick, this was frugal fare by any standards. The parent no doubt tried to compensate by bringing back as many fish as possible on one trip, judging by how many of these fingerlings it had endeavoured to stack in its beak. At the other extreme, if a Puffin chances on a shoal of well-grown sprats, it may only be able to stow one or two at once in its beak, but these will make every bit as good a meal for the chick as scores of more inferior fish.

Of course, Puffin parents need to look after their own energy needs as well as those of their chicks, so not all the fish they catch are kept for transporting back to the colony. On their first foraging trip in the morning, Puffins probably satisfy their own needs first before thinking of the 'children'. Based on modelling data from diving performance, chick-feeding rates and energetics, it seems that Isle of May Puffins caught a fish in every one dive out of two. They also appeared to eat most of what they caught, retaining less than one in ten fish (sandeels) to bring back to their chicks. Possibly they have a tendency (as terns do) to eat the smaller prey themselves and keep the bigger prey for their chicks. A Dutch study of captive free-swimming Puffins found that when they were offered a choice of different fish species of varying sizes, they swallowed small fish underwater (sometimes several in the course of one dive), but brought the larger fish to the surface.

This begs the question of how Puffins get rid of the salty seawater they must inevitably swallow when they consume fish underwater. The salt they take in is absorbed into the bloodstream and partly excreted through the kidneys. However, the surplus, as in other ocean-going seabirds, is stored in a pair of well-developed internal salt glands, one above each eye. From this, a concentrated salty fluid is secreted from the nostrils (a horizontal slit near the base of each side of the upper mandible) and trickles down the bill. From time to time the Puffin simply shakes these drops off.

In a colony it can be hard to look beyond the whimsy of the Puffin's parrot bill and its dinner-suited apparel. But more than what meets the eye of the Puffin's anatomy and behaviour has been shaped over millennia by the demands of deep-pursuit diving for fish in an extremely challenging environment, the open ocean. This knowledge tempers our affection for the bird with genuine respect and admiration.

Above: On the Isle of May, a couple of big sprats make for a substantial meal.

Below: A 20cm sprat caught in the waters off Skomer Island and destined for some well-grown youngster.

OSRL

Threats to Puffindom

The Puffin is a hardy, resilient bird but it faces numerous dangers on land and at sea. Breeding in an underground bunker provides safety, but also invites the hazard of subsidence and flooding. Apart from such acts of God, all stages of Puffin breeding are at risk of being gatecrashed by other species, and beakfuls of shiny fish are like a neon sushi bar to a variety of bird-food pirates. Some of these have an appetite for the Puffin itself, as do some Nordic human communities that regard the bird as a delicacy.

Beyond the colony, Puffins also face threats from oil pollution, commercial fishing and the insidious impact of climate change. Indeed, puffindom is under assault as never before from the mounting pressures we humans exert on the oceans.

If you had asked anyone a few decades ago about the biggest threat facing Puffins, most would have said 'oil pollution'. Foremost in their mind would have been a litany of oil-tanker disasters, beginning with the shipwreck

Opposite: The oil tanker *Prestige* inflicted terrible damage on the Galician coast and its seabirds.

Below: Puffin victims of the 20 million gallons of fuel oil spilled from the *Prestige*.

Above: Volunteers scour the stricken Pembrokeshire coast for oiled seabirds after the *Sea Empress* ran aground.

Top: The oil tanker *Braer* foundered on Shetland's Sumburgh Head in 1993.

of the *Torrey Canyon* on the Cornish coast in 1967, made all the more notorious for the vessel being bombed by the RAF in efforts to burn off the oil, a folly never repeated since. This was followed by three tankers foundering in 1978, then by the *Braer* (Shetland, 1993) and *Sea Empress* (South Wales, 1996). Thankfully there have been no more shipwrecks since the *Prestige* broke up off Galicia in 2002, although this disaster hit Puffins hard. Out of over 23,000 oiled seabirds collected on Spanish beaches, 3,854 were Puffins, including 22 ringed birds. Apart from one ringed in Wales, all the rest hailed from Scottish colonies, with no fewer than 17 coming from Sule Skerry in Orkney.

Distressing pictures of oil-coated Puffins floundering ashore, and the subsequent attempts to nurse them back to health, generate massive public awareness. However, while the suffering and death of individual victims are undeniable, there is little evidence (with the possible exception of Puffins at the southern edge of their range in the Channel Islands and Brittany) that any major oil-pollution incident has struck a mortal blow to any Puffin population.

A more chronic source of oil pollution at sea arises from vessels illegally flushing out their tanks. Legislation introduced to combat this is making a difference, with 'beached bird surveys' lately showing a decline in numbers of oiled Puffins and other seabirds washed ashore. Puffins still have to contend, however, with an ever more exotic cocktail of man-made chemicals at sea. This was dramatically illustrated in 2013 when over 4,000 seabirds, including Puffins, were washed ashore (many beyond saving) along the south-west coast of England, their plumage congealed in polyisobutylene, or PIB. This is widely used to thicken lubricants, and to make chewing gum, adhesive tape, lipstick and many other 'innocent' trappings of modern life. But 'innocence' comes at a cost: transported as a liquid in ships' tanks, PIB turns into a sticky glue (think chewing gum) in contact with seawater. With major public and political support, the RSPB, along with other NGOs, campaigned successfully to make it illegal for vessels carrying PIB to wash out their tanks offshore: a global ban on the deliberate discharge of high viscosity PIBs started in 2014.

Burrow blues

Above: Soft ground, well-worked by Puffins, faces the risk of burrow collapse.

Puffins have sometimes been known to burrow themselves out of house and home. A dense colony is a honeycomb of tunnels, typically in soft, peaty soil, and the constant use of the surface as a landing strip erodes its protective mat of vegetation. Bare patches develop and join up until the slope resembles a well-worked Rabbit warren. The advent of winter rains, not to mention wallowing seals in flatter areas, can lead to wholesale burrow collapse and, in the worst-case scenario, landslips that remove the entire carapace of soil. Faced with this ultimate catastrophe, a colony becomes uninhabitable and the Puffins are forced to literally seek pastures new.

A number of Puffin islands have suffered this total loss of their core habitat. Once a thriving colony estimated in the 19th century at half a million birds, Grassholm's (West Wales) fragile cap became so riddled with burrows that it disintegrated, leaving a graveyard of tussocky stumps. By the 1920s only a couple of hundred Puffins clung on, but eventually even they lost their foothold and abandoned ship. Nowadays Grassholm is dominated by 40,000 pairs

Above: Grassholm has lost its Puffins, but Gannets prosper, coating the island like icing.

of Northern Gannets, but conspicuously no Puffins. While the exodus of Puffins from Grassholm appears on the face of it to be terminal, islands have been known to recover in time, especially with a helping hand from man (see pages 108–113).

Burrows are also prone to flooding, a hazard likely to increase in frequency and severity as climate change whips up ever more extreme weather events. In 2012, the wettest UK summer on record, Puffin colonies suffered disastrous basement flooding. On Brownsman Island alone in the Farnes archipelago, around 8,000 chicks or eggs were lost, with eggs bobbing around in burrows. In Scotland, flooding in the same season greatly reduced fledging success on the Isle of May.

Even without flooding, burrows are not always a secure vault. On Skomer Island, for example, around 6,000 pairs of Puffins co-exist with 120,000 pairs of another burrow nester, the Manx Shearwater. The Puffins do not compete for space with all the shearwaters as the latter nest mainly in the interior of the island, while Puffins favour

the cliff-tops. Even so, where the two species overlap, turf wars break out over burrows; with neither species seeming to have the upper hand, either may find itself evicted or have its egg destroyed in the melee.

Especially when an incubating adult is spooked, a Puffin egg sometimes gets accidentally kicked to the burrow entrance and snaffled by a gull that has been patrolling the colony, peering menacingly into burrows. Fortunately (or rather by design), the length and narrowness of most burrows makes the Puffin's nest chamber impenetrable to gulls. On Skomer, however, the inner sanctum is no barrier to the Jackdaw, a wily Gollum that itself nests in disused Rabbit burrows and is an arch-robber of Puffin eggs and fish. Unlike a bulky gull, a Jackdaw can readily insinuate itself into an unattended burrow, and roll out and crack open the egg with its beak. Exceptionally, a Jackdaw pair will even work as a hit squad on occupied burrows – one lures out the incubating Puffin while the other contrives to nip in and steal the egg.

Above: Burrow wars between the Manx Shearwater and the Puffin may go either way.

Below: The opportunist Jackdaw is partial to Puffin eggs and even fish.

Piracy and puffincide

As a group, seabirds have evolved two strategies for bringing seafood back to their young. Some are 'internal transporters', storing what they catch (including tiny zooplankton) in their crop. The other group, including all the puffin species, are 'external transporters', almost exclusively of fish. A major downside of this strategy is that your catch is exposed for all to see. In the case of our Atlantic Puffin, a beakful of silvery fish held crosswise in the beak is an open temptation to marauding gulls and other so-called 'kleptoparasitic' species.

The gull species (Herring, Lesser Black-backed, Great Black-backed, Black-headed) that co-habit with Puffins on UK islands all indulge in robbing Puffins of their fish as they arrive at the colony, and some individuals specialise in such piracy. Alert to the danger of being ambushed in the air or on the ground, a fish-carrying Puffin will typically circle its sub-colony area, judging the right moment to make an unerring dash for home.

This goes a long way to explaining why Puffins prefer to nest around the perimeter of islands on slopes directly exposed to the sea, rather than in the flat interior which demands running a longer gauntlet. David Nettleship's study of Puffins on Great Island, Newfoundland, found

Below: A Farne Islands Puffin struggles to deny a Black-headed Gull robber.

that the birds nesting on the flat lost their fish loads to Herring Gulls more often than those on the slopes. Piracy was rampant, with almost a third of incoming Puffins being routinely attacked and one in ten losing their hard-won meal to the gulls.

Gulls, and more often skuas, will harry and buffet Puffins in the air to try and make them drop their fish, which they then retrieve. By flying flat out and doing its best to jink, a Puffin may frustrate its assailant, especially where the protagonist is a gull. On the ground, however, gulls are more successful, pouncing on and grabbing hold of Puffins. In desperate efforts to hang on to its precious meal, a victim will stand its ground and fight back, beak and claw, sometimes emerging the winner. Where the attacker is a powerful, large gull, the Puffin may find itself seized and carried aloft.

Forfeiting meals to gulls can seriously affect chick rearing, but the setback pales in comparison to the Puffin losing its life. Puffins have been recorded as prey in the diet of the falcons and eagles that patrol sea cliffs, especially in Arctic regions. Some Great Skuas can also make heavy inroads into Puffin colonies, terrorising burrows and airspace in equal measure, but the number of UK colonies at risk is limited to Scotland's Northern and Western Isles.

By contrast, the Great Black-backed Gull, the biggest of our native gull species, is a widespread Puffin killer.

Above: Seizing a Puffin in the air, the Herring Gull will panic it into forfeiting its meal.

Below: Puffin chicks not uncommonly fall victim to the Great Skua (or 'Bonxie').

Above: Some Great Black-backed Gulls have perfected the art of plucking Puffins out of the air.

A voracious predator, scavenger and food pirate, this omnivorous gull enjoys a diet ranging from flying ants to Rabbits. It was on a visit to St Kilda in June 1971 that I first appreciated the Great Black-backed Gull's appetite for Puffins. Skins (neatly turned inside out) of eviscerated Puffins littered the nesting territories of certain pairs like tribal fetishes. These gruesome middens also contained remains of gull and skua chicks, along with carcasses of adult Razorbills, Kittiwakes and Oystercatchers.

Great Black-backed Gulls employ a variety of Puffin-hunting tactics. In one ground-launched assault I witnessed, the gull stood sentinel just above a burrow and grabbed the Puffin by the neck as it emerged. The victim struggled violently and, as the gull tried to fly upwards with it, managed to wriggle free and make good its escape, apparently unharmed. On other occasions, announced by an audible crack of primaries flexing in the air, a predator would swoop and pluck a Puffin out of a revolving wheel (see page 41). The outcome was not always fatal, however; sometimes in a desperate last throw of the dice the Puffin body-swerved, leaving the gull to overshoot into empty space, or else the Puffin plummeted to the water, typically jettisoning its meal as it did so (perhaps a decoy tactic to distract the gull from the bigger prize).

On St Kilda's island of Dun, Kenny Taylor discovered that some Great Black-backed Gull pairs staged a cunning double act, the better to ambush Puffins unawares: one member of the pair would tower high

Below: Against all the odds, this juvenile Northern Gannet in the Firth of Forth escaped the jaws of a determined mink.

A fishy tale

More danger lurks for seabirds beneath the waves than is often realised. In May 2011 Michael Patten was fishing with a rod and line north of Lappa Island, Norway, when he landed a handsome cod weighing nearly 12kg (26lb). Opening its bulging stomach, he was astonished to find a 500g (1lb) Puffin that had clearly not long been gulped down. Cod is a top predator among fish and, while other fish species are the mainstay of its diet, it is an opportunist and will devour whatever comes within range of its jaws, in this case an unsuspecting Puffin (the angler noted numerous Puffins diving in the area at the time).

There are rare reports of other diving seabirds – and inexplicably also a partridge! – found in cod stomachs, but this appears to be the first record (documented by Declan Quigley in *British Birds*, April 2012) of a Puffin. Other predatory fish pose an occasional hazard to diving auks: 14 Little Auks were found in the stomachs of monkfish (also known as anglerfish) caught in gill-nets off Cape Cod between 2007 and 2010.

Above: Gutting a cod led to the astonishing discovery of a taste for Puffin.

above the colony while its mate flew fast and low over the slopes, calling loudly. Engulfed by this advancing 'bow-wave of dread', perched Puffins either scuttled down burrows or else were flushed into flight. As Puffins panicked upwards, the pincer movement climaxed, the ground-sweeper's soaring accomplice stooping to single out and strike a bird caught up in the mayhem.

While other birds pose the main threat to Puffins, mammalian predators also take their toll. By eating their eggs and small chicks too, rats (both the Brown and Black species) have been responsible for decimating and even exterminating some Puffin colonies. Feral Mink can also wreak havoc and may well have extinguished the small Swedish population of Puffins some 50 years ago. Given that Mink can swim without stopping for three hours, a Puffin colony probably needs to be separated from the nearest population of this amphibious predator by several kilometres of open water to guarantee immunity.

Below: A Lesser Black-backed Gull mugs a Puffin offshore from Skomer Island.

Puffin on a plate

On remote rocky islands, ill-equipped to feed human communities by other means, man has traditionally harvested Puffins and other seabirds for food. This was a tradition on both sides of the Atlantic, with the native communities of Labrador and Maine as partial to Puffin as those in Norway, Iceland, the Faroes and Scotland. Puffin is no longer on the menu in Scotland and Norway, but Icelanders and the Faroese still have a taste for it. In Iceland the tourist is offered smoked Puffin by hotels and restaurants as a national delicacy, echoing times past when it was a vital source of protein and a welcome alternative to cured fish.

When the small community of souls from Hirta village on St Kilda, 60km (37 miles) west of the Outer Hebrides, was finally evacuated to mainland Scotland in 1930, there died out an ancient culture and skill set of dogged self-sufficiency unique in Britain. Fortunately St Kilda's daily grind of subsistence was well documented by a succession of fascinated, sometimes voyeuristic visitors and clerics. Bountifully blessed with a host of seabird species, St Kilda's Puffins, Fulmars and Northern Gannets were lifelines to the community, which turned to good use everything these birds could offer, whether meat, eggs, feather down, oil for lamps and even stomachs to fashion into bags and pouches. St Kilda's men were intrepid climbers, weaned from childhood on scaling the islands' towering sea cliffs barefoot or wearing thick socks. They used a variety of Puffin-catching methods, variously horsehair snares, or a pole topped with either a noose or a net.

Below: A successful 'sky fishing' attempt by an Icelandic hunter.

Perhaps an itinerant seafarer introduced the netting technique to St Kilda from the Faroes, origin of a centuries-old fowling tool called a 'fleyg', or 'fleygastong', consisting of a triangular net slung between two thin arms mounted on a 3m (10ft), hand-held pole. The fowler hunkers down on the cliff-top at a spot strategically placed to intercept the Puffins wheeling just overhead. Spying a bird approaching, the fowler sweeps his fleyg upwards to try and pluck the bird out of the air. In the heyday of such 'sky fishing', an experienced hunter could easily catch and kill a hundred or more Puffins in a day. There are archival photographs of Faroese fowlers at the turn of the 20th century with garlands of Puffins strung like trophies around their waists.

It was in the community's interests, however, not to cull more than the Puffin population could sustain. So the Faroese restricted the length of the hunting season (July–August) and the duration of the hunting day, and sometimes set a maximum quota. They also avoided catching fish-carrying birds in efforts to selectively kill immature birds (less vital to colony survival) rather than adults feeding young; starting the season at the beginning of July made it more likely to catch immature Puffins prospecting the colony late in the season than the breeding adults.

A hundred years ago the Faroes supported a healthy two million or so pairs of Puffins, capable of supporting an annual fleyg tally of hundreds of thousands. By 2000 the catch had dropped to a still-substantial 95,000, but by 2010 the hunt yielded just over 300 Puffins. This staggering downturn reflects a four-fifths decline in breeding numbers since the halcyon days of Faroese puffindom, and hardly any pufflings have fledged for the past 10 years. In response to this crisis, nearly all landowners have belatedly agreed a hunting ban on their colonies. Before we rush to judgement, however, on the impact of hunting, it is instructive to look at the strong echoes of the demise of the Faroes' Puffins in Iceland, some 700km (435 miles) to the north, and consider whether deeper forces might be at play here.

Above: A day's haul of Puffins on Iceland's Westmann Islands.

Below: In the heyday of Icelandic hunting, Puffins could be strung as a necklace for taking home.

Iceland is by far and away the Puffin capital of the North Atlantic, supporting around 2.5–3 million pairs. In 1994 it was decreed that hunting with fleygs could only take place from 1 July to 15 August. Puffins arrive in Iceland in late April and leave at the end of August, so the open season for hunting was chosen to coincide with the maximum presence of the late-arriving immature birds to relieve hunting pressure on the May–June breeders. Thus non-breeding two-to four-year-olds have traditionally been claimed to comprise over 90 per cent of the catch, although nowadays many more nesting birds are allegedly killed in the hunt.

As on the Faroes, Iceland's Puffins have declined catastrophically in recent years, with colonies in the south and west hardest hit. Records of the Puffin harvest reflect this: in the mid-1990s over 200,000 birds were caught annually in Iceland (by 100–200 hunters), compared with fewer than 40,000 in recent years. Some conservationists argue that with the population in such dire straits, harvesting Puffins is unsustainable and indefensible, and that a complete ban should be imposed. Iceland's environment minister has the power to restrict or even ban hunting nationally or regionally, but has never enacted this. In practice the power lies with each island's local authority, effectively landowners, who in turn lease Puffin hunting to individual clubs, so each island holds sway over whether to hunt or not.

The Breiðafjörður islanders, for example, have chosen to stop hunting altogether. However, arguments rumble on in the Westmann Islands, the heartland of Iceland's Puffin population with upwards of a million pairs, where the hunting culture is deeply embedded and jealously guarded. Since 2008 the response to ghost-town Puffin colonies has fluctuated annually between either a curtailed hunting season or else an outright ban. In 2013 the Westmann Islands council allowed a token five days (19–23 July) of hunting 'so this tradition does not become lost', but so few Puffins were around to be caught that it was scarcely worth the effort. One hunter netted just 16 birds, complaining that an unfavourable wind direction for fleyging compounded the dearth of young birds.

Hounding Puffins

An ingenious way of catching Puffins was devised on the Norwegian Lofoten Islands, a mountainous archipelago and Puffin stronghold. Here evolved the *lundehund*, literally meaning Puffin dog, a breed known from the Middle Ages. Our own Puffin island of Lundy in the Bristol Channel is a reminder of this Norse link. The dog looks a bit like a small Husky or a Corgi, but there the comparison ends because of its unique anatomical adaptations for seeking out and retrieving pufflings underground at night.

Each of the Puffin dog's paws has six muscular, jointed toes (all other dogs have four), the better to gain purchase on the Lofotens' slippery, boulder-strewn slopes. Indeed all of this dog's joints are remarkably flexible, vital for squirming and contorting in the twisting labyrinth of Puffin burrows. Here is a universal-joint dog that can bend its head backwards over its shoulders, and its legs flat against its body in almost any direction, even straight out sideways if needs be. Once it worms its way down a Puffin burrow, it can also close its ear canals at will to prevent soil getting in, and its thick coat is equally dirt resistant.

On a good night a Puffin dog could catch over 100 pufflings for its owner, and an experienced retriever was as prized as a milk cow. Distemper nearly wiped out the breed in the 1940s but it was rescued from the brink and, with Puffin hunting by dogs (or indeed any other method) now outlawed in Norway, the breed is sought after today as a household pet.

Above: The Norwegian Puffin dog, purpose-built for hunting underground.

Our simmering oceans

While hunting arguably adds to the pressure on the Nordic Puffin populations, it cannot be held responsible for the Puffin crash in the first place. Something else is happening and, as is the case so often with seabird populations, it is food shortage. It is clear that sandeels are disappearing from these waters and Puffins have no ready alternative prey to turn to. There are strong suspicions that the Puffin collapse in Iceland and the Faroes is associated with large-scale oceanographic changes in the North Atlantic, allowing warm subtropical water to surge northwards. This in turn appears to be disrupting the entire food chain, depressing the abundance of cold-water plankton and the cold-water sandeels that feed on this 'primary production', and ultimately depriving Puffins and other

Above: Sandeel shoals are dwindling as our seas warm up.

M Edwards, SAHFOS

Above: The cold-water plankton *Calanus finmarchicus* once propped up the food chain but is fast disappearing from our waters.

sandeel-dependent seabirds of food. The impact has been so dramatic on the Faroes that since 2003 breeding has collapsed across the board for Puffins, Arctic Terns and Black-legged Kittiwakes.

However, this is not just a Nordic saga – it is affecting northern UK Puffin colonies as well. The state of puffindom in the UK up until about 10 years ago was one of population increase over several decades. Since then, however, the pendulum has swung dramatically the other way. The situation is most dire in Orkney and Shetland, which geographically have as much in common with the Faroes (about the same distance north of Shetland as Durham is from London) as they do with mainland Britain. Sandeels are disappearing from these Northern Isles and the Puffins are experiencing breeding failure year on year, as are all the other seabirds there whose staple diet is sandeel. Chapter 'From Egg to Puffling' touched on the breeding struggles exposed by the webcam at Sumburgh Head, in microcosm a malfunction afflicting all of Shetland's colonies.

Climate-driven sea temperatures are now higher in UK waters than at any time since records began. Since 1870 the annual sea-surface temperature has increased by up to 1°C (1.8°F), which seems a modest increase, but for a marine ecosystem it is a massive cranking up of the heat. In the North Sea it has triggered a 'regime shift' in which the overall abundance ('biomass') of planktonic copepods has declined by a staggering 70 per cent in the last 50 years. The cold-water copepod has progressively been replaced by a warm-water species that is not only much less abundant but also has a different seasonal cycle, out of kilter with the emergence of plankton-eating sandeel larvae. As a result the average productivity, length, weight and nutritional value of North Sea sandeels have declined markedly over the last decade.

Faced with this squeeze on their staple diet, Puffins are forced to hunt for poorer quality and sometimes indigestible fish at greater distance from the colony (see page 85). This puts a huge strain on Puffins throughout their breeding cycle: adults arrive back in the colonies in poorer condition and often late, fewer of them lay eggs and those that do may lay smaller eggs (see page 59). With the deficit mounting, chicks starve, fewer survive to fledging and those that do are often underweight. In 2005 on St Kilda, for example, average fledging weight was only 157g (5½oz) compared with 250g (9oz) in previous healthy years. Mike Harris and Sarah Wanless show that this is part of a widespread, long-term decline since the 1970s, mirrored by pufflings also fledging progressively lighter year on year on the Isle of May and apparently also Sule Skerry (an island west of Orkney). The survival chances of underweight young are stacked against them when they leave the colony. As long-lived birds, Puffins can ride out the odd season where they draw a breeding blank, but years of recruitment failure are now taking their toll on population numbers.

To add insult to injury, Puffins are encountering increasingly stormy sea conditions (another predicted symptom of climate change) outside the breeding season. In March 2013, when Puffins were returning to their North Sea colonies, a prolonged spell of easterly gales resulted in the worst Puffin 'wreck' in at least 60 years, with 3,500 emaciated corpses beached all along the east coast and doubtless many more unaccounted for offshore. The scouring wind churned up the water column, severely handicapping the Puffins' ability to feed – in the words of Mike Harris, it 'knocked the stuffing out of them'.

For reasons we do not yet understand, however, the impact of climate change is not evenly spread across the UK. Perhaps counter-intuitively, Puffin colonies further south in the North Sea, and likewise the Irish Sea, seem to be faring better than those in the far north. Thus while colonies on the Isle of May and the Farne Islands each lost about 20,000 pairs between 2003 and 2009, the next census in 2013 showed no further deterioration and the Farnes population even increased a little. Puffins in Orkney and Shetland may be especially disadvantaged because, unlike further south, they have no sprats as back-up prey when sandeels fail. Suspicions that Shetland's inshore sandeel fishery had contributed to the seabirds' woes in the 1990s were later dismissed, and the fishery itself has since become commercially extinct with the failure to find enough sandeels to make them worth fishing for.

Below: Victims of the North Sea Puffin 'wreck' of 2013.

Spare the sandeel

Above: Northern Gannets flock to a Danish sandeel trawler off the Firth of Forth before the fishery was excluded from the region.

Below: Denmark processes sandeels into fish meal and fish oil on an industrial scale.

The case of the Shetland's inshore sandeel fishery raises the issue of whether the much bigger offshore industrial fishery for sandeels in the North Sea poses a threat to Puffins. The fishery is confined largely to the Danish trawler fleet in summer (April–July). It is not run for human consumption but for processing sandeels into fishmeal and oil to make pellets, nowadays largely for rearing farmed salmon. In the 1990s the RSPB campaigned against the profligacy of allowing an annual catch limit as high as one million tonnes of sandeels, and also opposed the fishery operating so close to major seabird colonies like those on the Isle of May. In 2000 the EU responded to the conservation arguments by declaring a large inshore area (a 'box' in fisheries management terms) east of Scotland and north-east England off limits to the sandeel fishery, followed by other environmental improvements in how the fishery was managed.

Independently of these new constraints, however, commercial sandeel catches started to nosedive after 2000, and hit rock bottom in 2004. This coincided with the worst breeding season in living memory for North Sea seabirds, with widespread starvation of young. While it is tempting to think that the Danish fleet had overfished sandeels in 2004, the science does not support this; rising sea temperatures were almost certainly undermining the sandeels' ability to regenerate. Nevertheless it was important that the sandeel fishery did not turn the screw further on an ailing stock, so the RSPB welcomed the EU's decision to halt that year's fishing, even though the shutdown was imposed frustratingly late in the season. It is also vital to maintain the North Sea sandeel box into the future and to create marine protected areas around the UK's other key inshore sandbanks that are the go-to sandeel larders for Puffins and a host of other seabirds.

With no prospect of a quick fix to climate change, Puffins face a hazardous future. There will be twists and turns, and the trend will not be relentlessly downwards

USA Puffins in trouble

Climate warming is now affecting Puffins across the whole of their global range. On the other side of the Atlantic, symptoms worryingly similar to those affecting our Puffins are emerging. In winter and spring 2013, scores of emaciated Puffins were washed ashore dead from Massachusetts to Bermuda. In colonies in Maine the body weight of adults and chicks is declining. Chick survival slumped in summer 2012 as the staple herring deserted the islands, forcing the adults to target butterfish, which were often too big and round for the chicks to swallow. Blaming rising water temperatures in the Gulf of Maine, Steve Kress, Director of the National Audubon Society's seabird restoration programme, said, 'We don't know how the Puffin will adapt to these changes – or if they'll adapt.'

Above: An American Laughing Gull wrestles with a butterfish too big to swallow.

or evenly spread. Nevertheless, the long-term outlook is not promising. Debbie Russell's model of the impacts of climate change on British seabirds makes the sobering prediction that by the end of this century, climatic conditions may eclipse Puffin colonies in England, Wales (and the Irish Sea generally) and even south-east Scotland, as the species' tolerance limits shrink inexorably northwards. As we have seen, conditions in the far north are already far from ideal so it is not as if Arctic latitudes offer some Puffin utopia. Warm-water fish, like anchovy, spreading north with climate change might possibly help compensate for the Puffin's sandeel vacuum some time in the future, but further speculation for a marine ecosystem in such flux is unwise. In the meantime, we must do everything we can to enhance the Puffin's existing habitats.

Below: Puffins face an uncertain future as sea warming disrupts their fish supply.

Conserving Puffins

To prosper, Puffins ask no more than a safe place to breed and a plentiful food supply within easy reach. Unfortunately, there are no quick fixes for reining in the greenhouse gases that now seem to cast a lengthening shadow over the health of the Puffin's North Atlantic food chain. As the last chapter highlighted, we can shore up Puffins as best we can by ensuring a seabird-friendly sandeel fishery and protecting the places where sandeels live at sea. Beyond that our most practical, hands-on efforts inevitably focus on ensuring that our offshore islands provide a benign environment, free from ground predators, for Puffins to colonise, survive in and raise their offspring.

The UK has a wonderful archipelago of potentially thriving seabird islands, too many of which have been blighted by rats. Puffins and other burrow nesters cannot co-exist with rats, and at best hang on in inaccessible corners or at worst are wiped out altogether. As custodian of the biggest reservoir of breeding seabirds in Europe, the UK has a moral responsibility to get these sites back into good health. The good news is that we are steadily clawing back and transforming them. In recent years remarkable progress has been made in the art of exterminating rats, enabling Puffins and other seabirds to reclaim ancestral breeding grounds. In the UK, Glasgow University's Bernie Zonfrillo took up this restoration challenge 20 years ago on Ailsa Craig in the Outer Firth of Clyde. After the island was cleared of rats, it took 10 years for the Puffins to re-establish themselves, and today over 100 birds are regularly seen on the island.

With variations dictated by the scale and topography of the target island, the basic approach is as follows: humane rat poison (harmless to seabirds) is distributed in a regular grid, either at bait stations or – on large islands – from the air by helicopter. To gauge the impact on the rats, lollipop-like 'chew-sticks' soaked in fat are then

Opposite: The fortunes of Puffins are highly dependent on a healthy food chain.

Below: Checking a rat bait station on Lundy Island.

stuck in the ground and inspected later for telltale rat tooth-marks that pinpoint hotspots in need of rebaiting. This is repeated until no sign remains and the island can be declared rat-free. The vital last piece in the jigsaw is to put quarantine measures in place to prevent any future reinvasion of stowaway rodents from boat landings, cargoes and even personal belongings – visitors have to be super-vigilant about what they bring onto islands.

After successful rat eradication, Puffins and other seabirds have returned in recent years to nest on Cardigan Island and Puffin Island in Anglesey, and Handa and most recently Canna in the Western Isles. Being near the southern limit of the Puffin's range in the UK, the restoration of Lundy in the Bristol Channel is especially noteworthy. Ten years after the island was declared rat-free, a survey of Lundy by the RSPB (in partnership with the National Trust, Landmark Trust and Natural England) in spring 2013 discovered that Puffins had increased from a precarious five birds to 80. Reduced to only 300 pairs in 2003, Lundy's Manx Shearwaters have since expanded ten-fold, making it England's most important site for this species, and other seabirds have also bounced back. Spurred by Lundy's success, the programme is now being extended to the Scilly Isles.

Hugely encouraging as the resurgence of Puffins is on these islands after rodent control, it may seem surprising not so see an even stronger recovery. Many factors come into play here. If the island has been long abandoned by Puffins, there are no home-bred survivors with any 'memory' of the site. Moreover, the distance to other nearby colonies and the status of the latter's own Puffin populations will affect the reoccupation rate. Last but not least, Puffins are, by nature, highly social birds and 'need' to be in a colony to feel safe, find mates and breed successfully. So an island devoid of Puffins, however tempting in other respects, may not necessarily trigger a stampede of nomadic, home-hunting Puffins.

An interesting case in point is Ramsey Island, an RSPB reserve off the Pembrokeshire coast. Puffins last bred there in 1894, so we know that it does at least have a history of occupation. Shipwrecked Brown Rats had been on Ramsey ever since the late 1800s until they were eradicated in the winter of 1999–2000. Other seabird species burgeoned, but so far Puffins – which flourish just a stone's throw away on Skokholm and Skomer – have resolutely failed to join them. The wardens, Greg and Lisa Morgan, resorted to the ploy, first developed in North America, of planting decoy Puffins on the island to give the impression, at least from a distance, of it being occupied, a 'social-attraction' method pioneered 40 years ago by Stephen Kress on Eastern Egg Rock Island in the Gulf of Maine (USA).

Like many Maine seabird colonies, Eastern Egg Rock's Puffins had been so decimated by a combination of historical hunting and rocketing gull populations that they ceased breeding there in 1885. Among the efforts to bring them back, Kress and his National Audubon Society team introduced two highly innovative approaches, firstly 'translocating' nearly 1,000 young Puffins from Newfoundland, and secondly (aware of how much Puffins like to interact socially) planting painted wooden decoy Puffins and mirror boxes on the island. Unlike decoys, translocation – the transplanting of potential settlers from another colony – is nowadays regarded as a last resort when all other restoration attempts have failed, and even then it should only be considered when the destination island is a guaranteed safe haven.

Above: Now that Ramsey Island is rat-free, Puffins are being given a helping hand to return.

SOS Puffin

Even if an island is rat-free, other factors may thwart the nesting prospects of Puffins. A classic case of this occurred on Craigleith in the Firth of Forth, where the once-flourishing Puffin colony, numbered at 28,000 pairs in 1999, suffered a dramatic crash to a few thousand birds. The Tree Mallow, a handsome salt-tolerant shrub that grows up to 3m (10ft) tall, was choking the island and turning burrow access into an obstacle course. The nearby Scottish Seabird Centre sponsored a five-year project, 'SOS Puffin', with squads of volunteers systematically shuttling out to the island in winter to chop out the menacing pest. The RSPB facilitated a parallel exercise on the neighbouring island of Fidra. With the Tree Mallow now firmly under control, Puffins have responded positively, returning in increasing numbers to breed on both islands.

Above: The colourful Tree Mallow is a Puffin barrier.

Above: Decoys on Ramsey Island help lure live Puffins ashore.

The mock-up Puffins exerted such a pull on Eastern Egg Rock that the first Puffins to return sidled up to the decoys and 'billed' with them (see page 53). Five pairs of Puffins began nesting in 1981, and by 2011 the island boasted well over 100 pairs. Returning to Ramsey Island, although 200 decoys have been deployed in potential burrow areas every year since 2009, as yet there has been no definitive proof of Puffins making landfall on the island. Nevertheless, numbers have built up on the water nearby, suggesting some degree of attraction. In 2013, to enhance the signal, a sound system to broadcast Puffin calls was added in the hope of replicating the impact of this refinement on Lighthouse Island in the Copeland Islands (Northern Ireland) where, after several months of using 'silent' decoys, the addition of calls resulted in 50 birds coming ashore within a week. On Ramsey several Puffins came ashore on low-tide rocks directly below the sound system, the first recorded landfall in 70 years. Some even ventured onto the slopes to check out the decoys, so all of these efforts may yet pay off.

Another variation on the decoy method has been tried to coax Tufted Puffins back to Hamanaka Island in Hokkaido, Japan. Since the 1960s the local Tufted Puffin population has crashed from several hundred pairs to around 10, none of which have actually bred on Hamanaka since 2008. The drowning of tens of thousands of the birds in gill-nets and drift-nets in Russian waters is almost certainly behind the collapse. Decoy Tufted Puffins were positioned not just on the island, but also at sea just offshore, using a floating modification. Tufted Puffins, probably of Russian origin, were certainly attracted to the floating decoys, but not enough to induce breeding on Hamanaka to date.

Below: Tufted Puffin decoys designed to float.

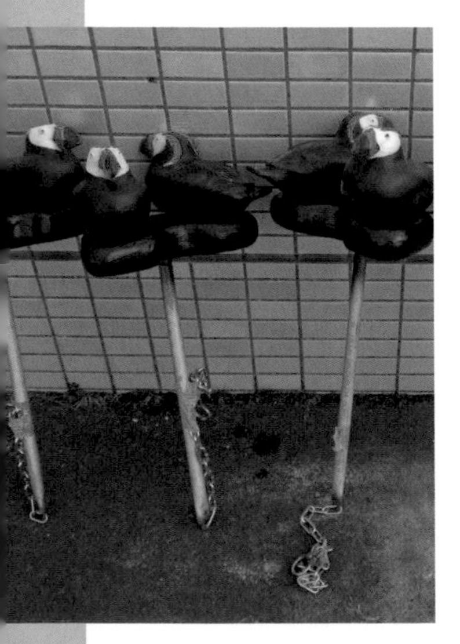

Common to all these conservation efforts is the basic need to count breeding Puffins. Because they are underground nesters, this is a challenge, especially in colonies where Puffins nest under rocks and in boulder fields. Ideally all the burrows (or a sample if this is impossible) are counted, then the proportion occupied is determined, preferably using an endoscope (see page 60), or failing that, the time-honoured and painful method

Puffin pioneers

Restoring Puffin islands is one thing, but once in a blue moon nature lends a huge helping hand by providing a brand new island. In November 1963 a local fishing crew noticed a column of smoke rising from the water off the south coast of Iceland's Westmann Islands. These were the birth pangs of the biggest submarine volcanic eruption in Iceland's recorded history. By June 1967 the outpouring of lava, pumice and ash had sculpted a brand-new island, christened Surtsey after the fire giant Surtr in Norse mythology. How flora and fauna would colonise this pristine natural laboratory was immediately the subject of intense international interest. To protect and study its evolution, Surtsey was made off limits to the public from the moment of its creation, and declared a UNESCO World Heritage Site in 2008.

By 1990, 20 species of plant had made it to Surtsey. Seabirds followed, first Black Guillemots and Northern Fulmars, and later various gull species. Breeding by Puffins was first suspected in 2001 and confirmed three years later, when a few were seen carrying beakfuls of fish into fissures in the sea cliffs. The terrain is still incredibly challenging for man and bird alike – rugged, crumbling lava with razor-sharp edges. The fragile topsoil, fertilised by the droppings of seabirds, will hopefully build up over time and improve the habitat for Puffins.

of extending arms down burrows. Occupied burrows can be marked for later inspection to judge the colony's breeding success. Census work is labour intensive and workers need to tread carefully to avoid making burrows collapse. However, it is a risk that has to be taken if we are to keep our finger on the pulse of how Puffins are faring nationwide and work out how best to look after them.

Left: Checking burrow occupation on Coquet Island with an endoscope.

Below: Numbered stakes identify breeding burrows on the Isle of May.

Puffin Fun

The Puffin is the most universally recognised and cherished of seabirds. This is largely because of its identikit appearance, which strikes many different chords in us – it is the rainbow-billed harlequin, the upright, dinner-suited dandy, the sad-eyed, orange-shod clown. Subliminal cuteness appeal, especially to children, may also lie in the Puffin's soft, rounded features.

As a socially interactive bird curious about anything novel it encounters, the Puffin again echoes the human condition. So Puffins seem to touch the sense of theatre, comedy and community in us all. This affinity also recommends the Puffin as the marine mascot of choice, a standard-bearer for raising public awareness about the need to safeguard seabirds and the oceans.

One way of measuring the affection in which we hold a bird is the readiness with which local communities invent a name for it. In the UK many of the Puffin's colloqulal names celebrate its beak – thus it is called 'sea parrot' and 'coulter neb' (literally 'ploughshare beak') in

Opposite: A Coquet Island Puffin is intrigued by its own image.

Below: Anything novel is readily explored.

A case of mistaken identity

In the RSPB's landlocked Loch Garten reserve in Inverness-shire, famous for restoring the Osprey as a breeding species in the UK, a visitor once allegedly reported to the warden a sighting of a Puffin in flight carrying a carrot. It was gently pointed out that, under the circumstances, this was more likely to have been an Oystercatcher.

Below: Entering head-first into the fun of the Amble Puffin Festival.

Northumberland, and 'nath' in North Cornwall, a Celtic name akin to the Welsh *nadd* – meaning something hewn or chipped off. Another old West Cornish name for Puffin is 'pope', or 'popey duck'. In Orkney and Shetland the Icelandic *nóri*, meaning the same as *nadd*, gave rise to the Puffin's endearing moniker of 'norrie', or 'tammie (tommy) norrie'. Even in its far-flung Mediterranean wintering grounds, the Puffin has entered the lore of Catalonian fishermen, going by the name of *Gavot de bec de Tomaquet*, the 'tomato-billed razorbill'.

The Puffin is *lundi* in Iceland and the Faroes (where, incidentally, the bespectacled Puffin cartoon character Ludvik Lundi is a household name), suggesting that our own island of Lundy echoes Viking empire building. Lundy's reputation as a haven for seabirds gained notoriety in 1929 when the island's purchaser, Martin Harman, issued stamps featuring a Puffin. Convinced that he was on to a good thing, Harman also minted coins – the 'Puffin' and the 'half Puffin' – with the bird on one face and his own profile on the other. However, the British government took a dim view of this self-appointed king of Lundy, fined him five pounds and prohibited the minting of the coins, despite Harman's protestations in court that he had sovereign rights. Luckily Lundy stamps were reprieved and are issued on Lundy to this day, while the original halfpenny pink and penny blue are sought after by philatelists.

Puffins are the star draw for tourists to Lundy and other islands, a magnetism which also makes them the centrepiece of a number of mainland festivals on both sides of the Atlantic. In Northumberland the Amble Puffin Festival was launched in summer 2013. Billed as celebrating 'everything Puffin', this family-orientated festival showcases community life in the small port of Amble, and also features boat trips to view Puffins and their young on the RSPB reserve of Coquet Island just offshore. During the Amble Festival an astonishing array of Puffin-related merchandise is testament to the bird's universal appeal – hats, fluffy toys, pin badges, cards and much more. Such products are not confined to Amble, however, but are used in coastal resorts and

Left: HRH Prince Charles, a stalwart of marine conservation, learns about Coquet Island's Puffins with the aid of a fluffy toy.

Below: Puffin feet as seen from inside a canvas hide.

Bottom: Puffins make ideal extras for conservation messages.

seabird visitor centres around the North Atlantic to fundraise for marine conservation through the lens of Puffin power.

On Coquet Island itself the Puffins are accustomed to the staff and their daily rounds, and routinely sit on the roofs of hides set up to study the nesting seabirds at close quarters. The innate inquisitiveness of the Puffin, more highly developed in this seabird than in any other I can think of, lends itself to interaction with human artefacts. This is evident from early on in a Puffin's life, as Kenny Taylor captures in his description of hand rearing pufflings indoors: 'The chicks had boundless curiosity, scuffling over to investigate objects on the floor, deftly picking up pencils, pecking at chequered markings on the linoleum or craning their heads to watch flies on the move'.

On Coquet the wardens have taken this one step further by putting a number of props at the disposal of loafing Puffins: a mirror, a pirate ship, an Olympics medal rostrum and even a swing. None of this harmless fun detracts from the Puffins' main purpose in life of reproducing themselves, but it often provides the imagery and metaphors we need to refresh serious stories about Puffins and the wider protection of the marine environment.

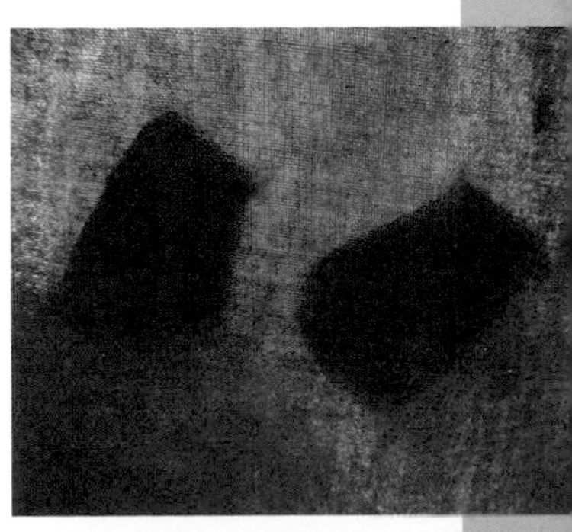

Above: Puffin appeal (left upper and lower) is captured by extrovert campaigning for marine protection (right upper and lower).

The RSPB has harnessed the Puffin to conservation efforts in many other ways. Since its launch in 1966, the society's *Birds* magazine (now called *Nature's Home*) has featured Puffins on the front covers of seven issues, more than any other species of bird. In Scotland staff donned the suits of Puffin, Northern Gannet and Black Guillemot (or 'Tystie') in the streets of Edinburgh to highlight the call for marine protected areas in Scottish waters (see photo above). In northern England a challenge to build a 'snowpuffin' helped to raise awareness of the RSPB's 'Safeguard our Sealife' campaign during the winter months. The RSPB also had an inflatable Puffin (see photo above) constructed for public events but it had a limited lifespan, having emerged from the factory with such King Kong dimensions that it risked taking flight in anything but the calmest conditions. The society's most recent innovation is the 'cappuffinccino', a coffee with a chocolate Puffin sprinkled on top, available in cafes and other outlets across the Yorkshire coastal town of Bridlington to celebrate the spring return of Puffins to the nearby Bempton Cliffs Reserve.

Such awareness-raising was actually pioneered 40 years ago on the other side of the Atlantic, when the National Audubon Society embarked on 'Project Puffin' to restore Puffins to their long-abandoned islands in the Gulf of Maine. The project has a number of sponsors committed to community action, including a domestic energy provider and a manufacturer of breakfast cereals called 'Puffins'. The sponsors' support includes school education programmes and internships, an adopt-a-Puffin scheme and a webcam. The cereal company also promotes 'the Travelling Puffin', in which visitors to the company's website are invited to buy a fluffy toy Puffin and photograph it beside world-famous tourist landmarks such as the Eiffel Tower and Colosseum, for posting on the website.

By incorporating a Puffin into its logo, the Maine energy company understands the value of hitching its wagon to 'brand Puffin', just as Allen Lane did when it launched Puffin Books for children at the start of the Second World War. The first Puffin logo (in this case depicting two Puffins nuzzling bills), however, was adopted by the Ligue pour la Protection des Oiseaux (LPO), the RSPB's French counterpart, which formed in 1912 to campaign against the safari-style shooting of Puffins in Brittany's Sept-Isles.

Author and naturalist Mark Cocker observed that: 'As long as we find ample living space for them, and no matter how we misinterpret or skew the facts, birds will continue to fly through our imaginations, dispensing a kind of wisdom as they go'. We are irresistibly drawn to what the Puffin communicates to us, and with a charm unique among seabirds. This in turn makes the Puffin a cornerstone for spreading human wisdom about the need for stewardship of its breeding sites and ocean habitat.

Top: Cup of Cappuffinccino!

Centre: Snowpuffin.

Above: Logo of the French bird conservation organisation 'LPO'.

Watching Puffins

Reading about Puffins is one thing, but a first face-to-face encounter in the breeding season is a spine-tingling experience. Sitting on top of the Puffin slopes on a breezy, sunny summer's day, all your senses are abuzz. You experience the din of seabirds shouting to be heard above one another; the air barrelling up the cliffs with a heady aroma of guano, salt, Thrift and Sea Campion, keeping Puffins and Fulmars dancing aloft as if for the sheer fun of it; the vertical sweep of colour from cobalt-blue sea to saffron lichen-encrusted rocks and back to blue again. Suddenly you feel twice as alive as you did back home.

So where is it best to experience this for yourself? A national breeding census of our seabirds is long overdue, but the last one, in 2000, mustered around 600,000 pairs of Puffins in Britain and Ireland, with the vast majority in Scotland, especially the Western Isles and Shetland. The Scottish mecca for Puffins is undoubtedly St Kilda with a population in excess of 140,000 pairs, but over 50,000 pairs may still reside on Sule Skerry, with Fair Isle and the Isle of May not far behind. The Flannan Isles, Foula and Hermaness also support major colonies.

Opposite: The Farne Islands has attracted generations of tourists determined to see Puffins.

Below: The distribution of the UK's Puffin colonies – the bigger the symbol the larger the colony.

Key: number of apparently occupied burrows

- 1 – 100
- 101 – 1000
- 1001 – 10000
- 10001 – 50000
- 50001 – 100000
- 100001 – 150000

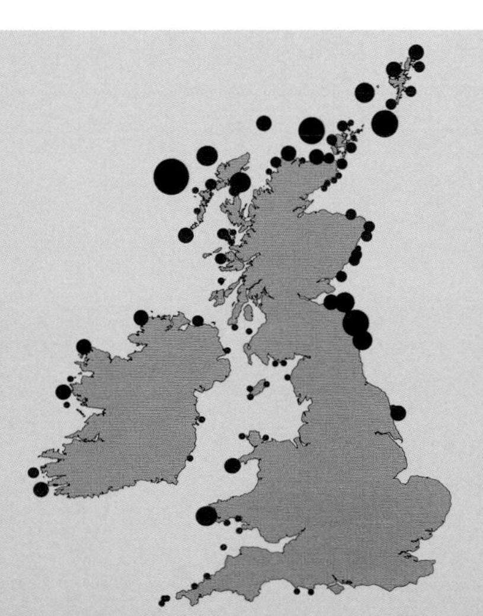

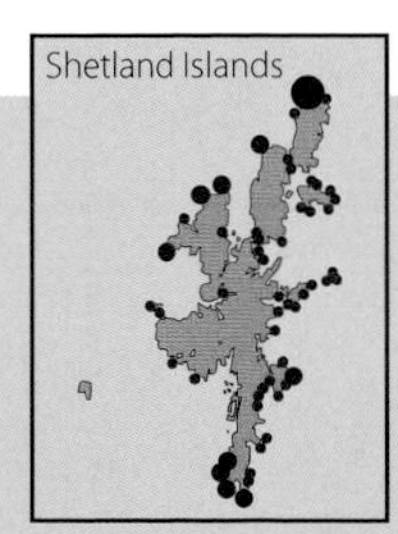

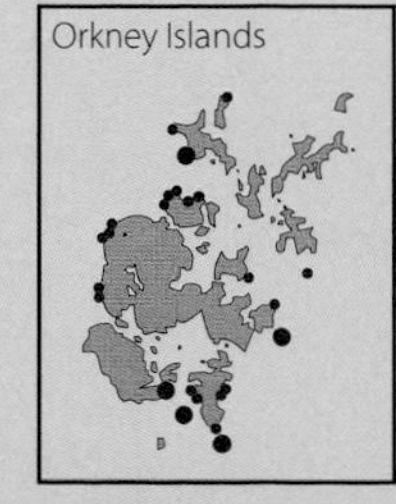

Above: Paths on Skomer Island prevent burrow trampling while providing the public with close encounters.

Below: The buoyant Irish Sea population of Puffins offers an excellent visitor experience on several islands.

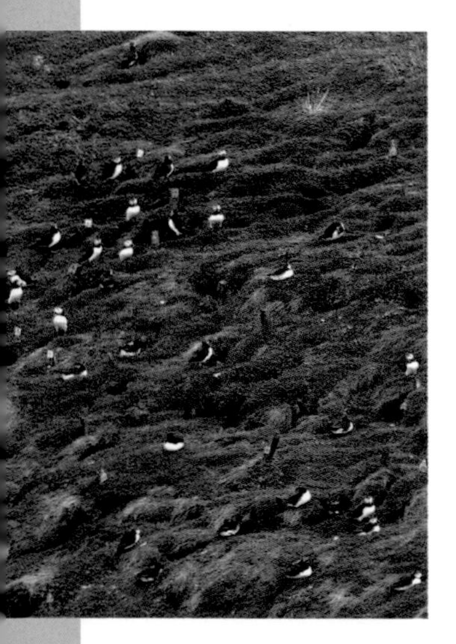

In England the Farne Islands in Northumberland now have a puffinry of around 40,000 pairs, with nearby Coquet Island home to rather less than half that number. There is also a significant colony in Bempton Cliffs (north Yorkshire), around a hundred pairs on the Isles of Scilly, and smaller outposts in Cornwall, Devon and Dorset, and on the Isle of Wight.

In Wales the heartland of puffindom is the Pembrokeshire islands of Skomer (7,000 pairs in 2000) and Skokholm (2,000 pairs). In Northern Ireland over 1,000 pairs breed on Rathlin Island (though many more do so on the west coast of Ireland). It is also worth mentioning the few hundred pairs of Puffins in the Channel Islands.

As most of these colonies are on islands, a visit requires forward planning and often the right sea conditions for embarking and landing. Some places cater only for day trips, but others have accommodation for longer stays, providing you book well ahead – all the key locations have associated websites. In Scotland you can stay on the Isle of May Bird Observatory or the Fair Isle Bird Observatory. For the ultimate seabird experience you could apply to join one of the highly sought-after summer work parties on St Kilda organised by the National Trust for Scotland. To stay on Lundy contact the Landmark Trust. For Skokholm and Skomer visit the website of the Wildlife Trust of South and West Wales. Remember that Puffins are only ashore from about April until August, and not all islands are open for business throughout that season. June and July, when nesting is in full swing, typically offer the best Puffin action.

For shorter trips numerous islands that host Puffins and other seabirds can be visited for a few hours (tides often dictate the times) or, failing that, circumnavigated by boat. Islands accessible for landing by boat (prebooked with the operator) include Ailsa Craig (from Girvan or Campbeltown), Handa (from Tarbet), Lunga in the Treshnish Islands (for example from Mull, Oban or Iona), the Farne Islands (from Seahouses) and Rathlin Island (from Ballycastle).

If landing is not on the menu, there are also guided 'cruises' (perhaps too grandiose a word in this context), which enable on-deck sightseeing of Puffins and much

Communing with Puffins

Whether you are on a managed excursion or Puffin watching on your own initiative, bear in mind that sea cliffs are dangerous places – this applies particularly to the grassy slopes that Puffins favour, so do take great care and use footwear with a good tread. Puffins are confiding birds and allow humans a much closer approach than do most seabirds, especially when they are 'loafing'. So hunker down in a safe place away from the edge and indulge yourself in what, on the island of Lunga in the Inner Hebrides, they like to call 'Puffin therapy'. Ian Morrison, who has spent his life ferrying tourists from Mull to Lunga, caters increasingly to visitors from all over the world who want to see a Puffin in the flesh. Talking to *The Scotsman* in August 2012, he said: 'I can see when they have gone ashore they are a wee bit apprehensive. It's as if I have dropped them off on a desert island. Then they come back two hours later and they have a big smile on their face, so Puffin therapy must work.' One of his passengers, a Chinese schoolteacher, said, 'It's absolutely true – the Puffins have a calming, relaxing effect on people. This was a trip I had been looking forward to for a long time.'

else, such as the possibility of dolphins and basking sharks, according to location. You can take such trips to the Inner Firth of Forth islands (*The Maid of Forth* sails from South Queensferry), Troup Head (departing from Macduff or Banff) and Coquet Island (from Amble: enquiries to the Amble Tourist Information Centre). Wherever there are Puffins, even in small numbers, it is always worth checking if there is a local operator or fisherman willing to offer a boat trip. Opportunities abound on the west coast of Scotland (the islands are dotted with numerous Puffin colonies not listed here), and also on the west coast of Ireland.

For those who lack sea legs, some Puffins are obliging mainland nesters and are highly accessible. Among mainland locations where they can be seen are the RSPB reserves at Sumburgh Head (Shetland), Dunnet Head (Caithness), Fowlsheugh (Aberdeenshire), Bempton Cliffs (north Yorkshire) and South Stack (Anglesey), all of which are wardened and offer great viewing points. The RSPB website can guide you to information on these and all its other seabird-colony reserves and how to get to them. In addition, Orkney and Shetland offer a host of possibilities for Puffins as part of an outstanding wildlife spectacle, whether from dry land, guided cruises or island landings.

Further details of these Puffin-watching opportunities are provided in the Resources section (see page 125).

Below: You're never too young to meet a Puffin (but mind that beak!).

Glossary

Bergmann's rule A principle, named after its discoverer, which states that populations and species of larger size are found in colder regions, while those of smaller size occur in warmer environments.

British Trust for Ornithology (BTO) Carries out research on British birds, chiefly by conducting population and breeding surveys and by bird ringing.

Centre for Ecology and Hydrology (CEH) The UK's centre of excellence for integrated research in terrestrial and freshwater ecosystems and their interaction with the atmosphere. The CEH carries out the world's longest-running population study of Puffins on the Isle of May in Scotland.

club area Site in the colony, often a flat rock, where Puffins congregate to rest, meet and interact with potential mates; also called a loafing area.

copepods Small marine crustaceans, many in the plankton mix in the water column (see also **zooplankton**).

decoy Mock-up Puffin deployed in numbers at an island's edge (less often in the sea) to attract Puffins to colonise the site.

endoscope Device (also known as a 'burrow-scope') with a miniature camera mounted on the end of a long, flexible probe for examining the contents and activity in a Puffin burrow (see also **Puffincam** and **Webcam**).

fleyg (or fleygastong) Net at the end of a long, hand-held pole, used for catching Puffins in the Faroes and Iceland.

geolocator Tiny electronic tag fitted to a Puffin leg ring that determines its flight path by logging how light levels change with longitude and latitude; the bird needs to be recaptured to download the data.

kilohertz Measure of the frequency of sound waves and thus the pitch of bird calls.

leucistic/leucism Reduced pigment in bird plumage, due to a genetic mutation.

Lundehund Norwegian breed of dog, adapted for hunting Puffins underground.

mate guarding Chaperoning by a male Puffin to prevent its mate from copulating with another.

moth flight Display in which a Puffin flies with outstretched, barely fluttering wings.

myoglobin Protein in the muscles of animals that stores oxygen, well developed in Puffins and other deep-diving seabirds.

Pelican walk Display in which a Puffin walks by raising each foot high in slow, exaggerated steps.

polyisobutylene (also 'polyisobutene', or PIB) Liquid, man-made chemical with widespread uses, which thickens in seawater to glue the plumage of seabirds.

Puffincam The internet transmission of an image from an **endoscope** (also called a **webcam**).

puffling Name for a Puffin chick from newly hatched to fledging.

raft Flock of Puffins on sea's surface near colony, featuring courtship and other social interaction.

radio tracking Means of recording the flight paths and offshore movements of seabirds by attaching various small devices that transmit location electronically.

regime shift Abrupt and fundamental, large-scale change in the marine environment and its food web.

selfish group (or selfish herd) Theory that predicts that a Puffin flocks with others of its own species to reduce its chance of being singled out by a predator.

time-depth recorder (TDR) Retrievable device attached to a Puffin leg ring, which records diving depth and frequency of dives over time.

webcam See **Puffincam**.

wheel Display in which Puffins simultaneously perform aerial circuits over the colony, apparently to regulate air traffic and for individual protection (see also **selfish group**).

wing loading The ratio of the weight of a bird (or an aircraft) to its wing area; the Puffin has a relatively high wing loading.

zooplankton Microscopic marine animals that, along with plants (phytoplankton) make up the ocean's floating plankton (see also **copepods**).

Further Reading and Resources

Not many books have been written about Puffins and some are out of print. The 'bible' is *The Puffin*, by Mike P. Harris and Sarah Wanless, illustrated by Keith Brockie (T & A D Poyser, 2011).

Sadly out of print is the excellent *Puffins*, by Kenny Taylor, illustrated by John Cox (Whittet Books, 1993), but you may be able to pick up a secondhand copy on the internet.

A detailed account of all aspects of the Puffin appears in vol. 4 of *The Birds of the Western Palearctic*, edited by S. Cramp (Oxford University Press, 1985).

For the most recent complete census of the British and Irish Puffin population, and interpretation of trends in numbers here and throughout the species' North Atlantic range, see *Seabird Populations of Britain and Ireland: Results of the Seabird 2000 Census (1998-2002)* by P. Ian Mitchell, Stephen F. Newton, Norman Ratcliffe and Timothy E. Dunn (T & A D Poyser, 2004). Preparations are underway for the follow-up national seabird census, the fourth of its kind.

For in-depth accounts of the Atlantic Puffin and its Pacific relatives, set in the context of the auk family, I recommend *The Auks: Alcidae* (Bird Families of the World) by A. Gaston and I. L. Jones (Oxford University Press, 1998).

Resources

This list of pointers should be read in tandem with 'Watching Puffins' on page 120. For details of each site, its location, access, what you will see and much else, go to www.rspb.org.uk/reserves/guide and enter the name of the reserve you want.

To get involved in the RSPB's Safeguard our Sealife campaign, contact www.rspb.org.uk/supporting/campaigns/sealife/copy_of_index.aspx

You can sponsor a Puffin for a year. For details see: http://shopping.rspb.org.uk/sponsor-a-puffin-for-a-year.html

The Marine Conservation Society (www.mcsuk.org) and The Wildlife Trusts (www.wildlifetrusts.org) are active in protecting our coasts and seas.

The Wildlife Trusts in England and Wales also manage a number of reserves where Puffins breed, notably:

www.ywt.org.uk/reserves/flamborough-cliffs-nature-reserve

www.welshwildlife.org/skomer-skokholm

www.wildlifetrusts.org/living-landscape/schemes/living-islands

The Scottish Wildlife Trust's reserves include Handa in the far north-west: http://scottishwildlifetrust.org.uk/visit/reserves

For the Isle of May: www.isleofmaybirdobs.org

For joining a work party on St Kilda: www.kilda.org.uk/work-party-information.htm

The National Trust also manages the highly accessible Farne Islands: www.nationaltrust.org.uk/farne-islands/wildlife

For Lundy: www.landmarktrust.org.uk/Search-and-book/staying-on-lundy

The Scottish Seabird Centre in North Berwick is much more than its Discovery Centre – it also offers opportunities to get involved in Puffin conservation and runs boat trips in the Firth of Forth: www.seabird.org/index.php

Puffins are also regularly seen on cruises around the Firth of Forth islands on *The Maid of Forth*: www.maidoftheforth.co.uk sightseeing-boat-trips-under-the-forth-bridge

Puffin cruises additionally run to a number of RSPB reserves, including:

Bempton Cliffs, North Yorks: www.rspb.org.uk/datewithnature/146968-gannet-and-puffin-cruises

Puffin Island, Anglesey: www.rspb.org.uk/datewithnature/149811-puffin-island-cruises-

Coquet Island, Northumberland: www.rspb.org.uk/reserves/guide/c/coquetisland

Troup Head, Aberdeenshire: www.puffincruises.com (from Macduff), www.northeastseaadventures.co.uk (from Banff)

Ailsa Craig, Ayrshire: www.rspb.org.uk/reserves/guide/a/ailsacraig/directions.aspx

Image credits

Bloomsbury Publishers would like to thank the following for providing photographs and for permission to produce copyright material. While every effort has been made to trace and acknowledge all copyright holders, we would like to apologise for any errors or omissions and invite readers to inform us so that corrections can be made in any future editions of the book.

Key: t=top; l=left; r=right; tl=top left; tc=top centre; tr=top right; cl=centre left; c=centre; cr=centre right; b=bottom; bl=bottom left; bc=bottom centre; br=bottom right

G = Getty; SH = Shutterstock; NPL = Nature Picture Library; RSPB = rspb-images.com

Front cover t Mircea BEZERGHEANU/SH, b Piotr Gatlik/SH; **spine** Piotr Gatlik/SH; **back cover** c Pim Leijen/SH, b Grant Glendinning/SH; **half title** Imagebroker/FLPA; **contents** Nigel Blake/RSPB; **4** Eric Isselee/SH; **5** Nigel Blake/RSPB; **6** Wes Davies; **7** tl Ingo Arndt/Minden Pictures/FLPA, tc Ingo Arndt/Minden Pictures/FLPA, tr Photo Researchers/FLPA; **8** Sheila Russell; **9** t Imagebroker/FLPA, c Wes Davies; **10** tl Sheila Russell, b Dave Boyle; **11** tr Wes Davies, b Barbara Fryer; **12** Keith Brockie; **13** Wes Davies; **14** t Mark Hamblin/RSPB, b Stephen Belcher/Minden Pictures/FLPA; **15** Neil Longhurst; **16** David Tipling/NPL; **17** Sheila Russell; **18** Laurie Campbell/RSPB; **19** tr Dave Boyle, b Sheila Russell; **20** tl Cyril Ruoso/Minden/FLPA, tr Dave Boyle; **21** Sheila Russell; **22** Neil Longhurst; **23** t Sheila Russell, br John Phillips/Wikimedia; **24** tl Graham Taylor/SH, b Mike Lane/RSPB; **25** tr Sheila Russell, b Mlenny Photography/G; **26** tr Sheila Russell, b Andrea Ricordi/RSPB; **27** t John Anderson; **28** bl Cyril Ruoso/Minden Pictures/FLPA, br Ben Lascelles; **29** Dave Boyle; **30** Doug Berndt/SH; **31** Paul Morrison; **32** tl Mike Jones/FLPA, b Mike Read/RSPB; **33** tr Dave Boyle, b Andrea Thompson Photography/G **34** tr Wes Davies, b Liz Mackley; **35** tr Wes Davies, c Rebecca Nason/FLPA, b Winfried Wisniewski/FLPA; **36** Tara Proud; **37** tr ImageBroker/Imagebroker/FLPA, c Ben Lascelles; **38** IMAGEBROKER, STEPHAN RECH/Imagebroker/FLPA; **39** Atlaspix/SH; **40** t Wez Davies, b Dave Boyle; **41** tr Steve Knell/RSPB, b Werner Bollmann/G; **42** t, b Dave Boyle; **43** t Dave Boyle, b JCluvsU2/SH; **44** t Neil Longhurst, bl Wes Davies, br Ben Lascelles; **45** t Grant Glendinning/SH, c Neil Longhurst, b Tara Proud; **46** Jules Cox/FLPA; **47** PavelSvoboda/SH; **48** tl Wez Davies, b Frans Lanting/FLPA; **49** tr Paul Morrison, b Tim Fitzharris/Minden Pictures/FLPA; **50** tl, tr, cl, cr Neil Longhurst; **51** c, b Dave Boyle; **52** Dave Boyle; **53** tl, tr, cl, cr Dave Boyle; **54** tl, tr, cl, cr Dave Boyle; **55** tr Wez Davies, b Neil Longhurst; **56** Richard Costin/FLPA; **57** Sheila Russell; **58** t David Tipling/RSPB, b Sheila Russell; **59** tr Mark Sisson/RSPB, br Eric Isselee/SH; **60** Gail Johnson/SH; **61** Neil Longhurst, br Annette Fayet; **62** Paul Morrison; **63** tl Dave Boyle, tr Wez Davies; **64** Mike Harris; **65** tr, b Wez Davies; **66** tl Sheila Russell, b Neil Longhurst; **67** Neil Longhurst; **68** Mark Medcalf/SH; **69** David Hosking/FLPA; **70** Adrian Ewart; **71** Dave Boyle; **72** t, b John Anderson; **73** Stuart Murray; **74** Neil Longhurst; **75** Paul Morrison; **76** t Wes Davies, bl Ben Lascelles; **77** t Wes Davies, bl, br Dave Boyle; **78** t, c Dave Boyle; **79** tr Bill Coster/FLPA, b Alex Mustard / 2020VISION/NPL; **80** Neil Longhurst; **81** Wes Davies; **82**

Jan Martin Will/SH; **83** bl Dave Boyle, br Sheila Russell; **84** Sheila Russell; **85** Steve Knell/RSPB; **87** c Mandy Lindeberg, NOAA/NMFS/AKFSC/Wikimedia, b Sheila Russell; **88** Neil Longhurst; **89** t Sheila Russell, br Dave Boyle; **90** Photo Researchers/FLPA; **91** Ricard Gutiérrez; **92** tl Eco Images/G, c Celtic Pic. Library/FLPA; **93** Sheila Russell; **94** Stuart Murray; **95** t David Tipling/RSPB, br Steve Trewhella/FLPA; **96** Neil Longhurst; **97** tr Dave Boyle, b Andrew Parkinson/FLPA; **98** t Andrew Parkinson/FLPA, b John Anderson; **99** t Declan Quigley, b Dave Boyle; **100** Jóhann Óli Hilmarsson; **101** tr Frans Lanting/FLPA, b Cyril Ruoso/Minden Pictures/FLPA; **103** t Gerard Lacz/FLPA, b Oxford Scientific/G; **104** M Edwards, SAHFOS, Sir Alister Hardy Foundation for Ocean Science, www.sahfos.ac.uk; **105** Mark Newell; **106** tl Chris Gomersall/RSPB, bl Euan Dunn; **107** tl Tom Vezo/Minden Pictures/FLPA; b David Tipling/FLPA; **108** Martin B Withers/FLPA; **109** Paul Glendell; **111** tr David Wootton/RSPB, br Robin Chittenden/FLPA; **112** tl Lisa Morgan, bl Yoshihiro Kataoka; **113** bl Paul Morrison, br Celine Marchbank; **114** Paul Morrison; **115** Wez Davies; **116** Wez Davies; **117** tr Paul Morrison, c Wez Davies, br Paul Morrison; **118** tl Wez Davies, tr Julia Harrison, cl Wez Davies, cr Ricky Devlin; **119** tr RSPB Bempton Cliffs, cr Rach Cartwright, LPO Alison Duncan; **120** David Tipling/FLPA; **121** Julie Dando, Fluke Art; **122** tl, bl David Hosking/FLPA; **123** Sheila Russell; **126** Neil Longhurst

Acknowledgements

Many thanks to those who supplied photographs, including some who made special efforts and trips to islands to match a shopping list of images we needed. In that regard, I owe a particular debt of gratitude to John Anderson, Dave Boyle, Wez Davies, Annette Fayet (and the Edward Grey Institute of Field Ornithology, Oxford University), Ben Lascelles, Neil Longhurst, Greg and Lisa Morgan, Paul Morrison and Sheila Russell.

Other photographs, Puffin facts and leads were kindly supplied by: Hólmfríður Arnardóttir, Helen Booker, Kara Brydson, Noelle Campbell, Carles Carboneras, Rach Cartwright, Keith Clarkson, Jess Crake, Rory Crawford, Gareth Cunningham, Heather Davison, Ricky Devlin, Alison Duncan, Adrian Ewart, Ricard Gutiérrez, Julia Harrison, Jóhann Óli Hilmarsson, Lucy Holmes, Mark Jessop, Yoshihiro Kataoka, Celine Marchbank, Crystal Maw, Helen Moncrieff, Stuart Murray, Mark Newell, Bergur Olsen, Ellie Owen, Chris Perrins, Ævar Petersen, Maria Prchlik, Tara Proud, Declan Quigley, Mayumi Sato, Scottish Seabird Centre, Mark Tasker and Alec Taylor.

I am grateful to Kenny Taylor for permission to quote from his own outstanding book on Puffins, and to Mike Harris and Sarah Wanless for their incomparable synthesis of all that is known to date about the species (see Further Reading, page 125, for both of these sources).

I have been greatly supported by Valerie Dunn, Derek Niemann, Jasmine Parker, (Senior Editor at Bloomsbury), Alice Ward and Krystyna Mayer's keen editorial eye.

Lastly I am grateful to all those who enabled me over the years to get to and work on seabird islands around the UK, including lighthouse keepers in times already fast receding into history. Visitors to these special places come away touched with something akin to what John Dyson attributed to those who work the sea: 'A fisherman has a look about him that makes the rest of us seem only half alive' (*Business in Great Waters*, Angus & Robertson, 1977).

Index